RECIRCULATING AQUACULTURE SYSTEMS

RECIRCULATING AQUACULTURE SYSTEMS

A GUIDE TO FARM DESIGN AND OPERATIONS

ANDY DAVISON

Recirculating Aquaculture Systems: A Guide to Farm Design and Operations by Andy Davison Published by Farmfish LLC Seattle, WA 2018

www.farmfish.org

© Farmfish LLC

All rights reserved. No portion of this book may be reproduced in any form without permission from the publisher, except as permitted by U.S. copyright law. For permissions contact:

farmfishbook@gmail.com

Cover by Chloe Kachscovsky

ISBN-13: 978-1-723-82344-2

Acknowledgements

I would like to thank all of my professors and mentors over the years for teaching me everything I know about being a writer, farmer, scientist, and engineer; my comrades in aquaculture engineering, Tomas and Maddi for sharing their time and knowledge (and homes); my partner Chloe for reading parts of my first draft and always supporting my passions; my parents Joel and Alexis for being nothing less than great parents every day; and to the rest of my friends and family for providing the network and support that every person deserves.

About the author

Andy Davison has a BS in aquatic and fishery sciences from the University of Washington and an MS in biological systems engineering from University of California- Davis. He has been working in the aquaculture industry for eight years in both the private and public sector. He has performed research on growing fish, shellfish, and seaweeds. This book is his initial project to synthesize some of the information he has learned over the first part of his career.

Table of Contents

Chapter 1: Introduction	11
Chapter 2: RAS Design Steps	18
Chapter 3: Production Plan	30
System Volume Calculation	33
Ammonia Production Calculation	34
Oxygen Consumption & Carbon Dioxide Production Calculation	34
Chapter 4: Water Quality	35
Chapter 5: Water Treatment Components	44
Tanks	45
Solids Filtration	50
Biofiltration	57
Carbon Dioxide Stripping	75
Disinfection	82
Oxygenation	89
Pumping	104
Fine Particle Filtration	117
Heating & Cooling	121
Alkalinity & pH Control	129
Chapter 6: Additional Equipment Considerations	132
Waste Treatment	132
Water Supply Treatment	135
Oxygen Generator	138
Ozone Generator	140
Emergency Oxygen	141
Fish Handling	144
Fish Processing & Transportation	147

Purging	149
Monitoring and Control System	150
Generator	156
Building	158
HVAC	160
Piping	162
Feeding System	167
Salt Mixers	172
Blowers	173
Lighting	174
Chapter 7: Operational Considerations	**176**
Cleaning	176
Monitoring & Water Testing	178
Fish Handling	179
System Maintenance & Corrosion	181
Purging	184
Feeding	186
Biosecurity	189
Disease Management & Treatment	193
Staffing	194
Chapter 8: Other Topics	**197**
Regulations & Permitting	197
Construction & Engineering Partners	198
Fundraising	200
Aquaponic Systems	204
Hydroponic Systems	207
Polyculture Systems	208
Biofloc Systems	209
Hatchery Systems	211
Shellfish Hatchery Systems	212

Algae Systems	214
Post-Smolt Salmon Systems	216
Seafood Holding Systems	217
Developing World RAS	219
Hobbyist RAS	220
Robotics	221
Chapter 9: Summary	**223**
Appendix A. Aquaculture Design Process	225
Appendix B. Farm Outline	226
Appendix C. List of Components for Costing Consideration	228
Appendix D. 100 Design and Operation Questions	230
Appendix E. 1000 MT Farm by the Numbers	235
Appendix F. Water Treatment Equipment Choices	236
Appendix G. Top Ten Most Common Mistakes	242
Appendix H. Business Plan Outline	246
Appendix I. Operations Checklist	248
Appendix J. How to be a Great Farmer	250
Appendix K. Additional Resources and Reading	252
Appendix L. List of System Designers and Equipment Providers	254

Chapter 1: Introduction

Chapter 1: Introduction

The purpose of this book is to provide a practical overview of recirculating aquaculture systems (RAS) that will assist in the construction and operation of these systems. The book is primarily targeted toward first time farmers or investors who are looking to begin a sustainable business, but hobby enthusiasts will also find useful information on the following pages. The book will provide an overview of all the essential information needed to begin making design and business decisions when planning a profitable RAS farm. Following Pareto's rule of efficiency this book intends to provide 80% of the information needed to be a competent recirculating aquaculture system designer and operator with 20% of the effort. As a result, this book is less concerned with the science and theory of RAS water treatment processes and more concerned with practical considerations. If you wish to go even more in depth into the understanding of recirculating aquaculture systems then reference the list of recommended resources at the end of this book. Reading these will get you another 15% of the way towards understanding recirculating aquaculture systems, and the last 5% will come from years of personal experience and conversations with experts.

RAS is defined as an aquaculture system that has a water exchange rate of 10% or less of the total system water volume per day. RAS comes in many shapes and sizes, all systems include basic components such as biofiltration, and their complexity depends on the goals of the system. Systems can

Chapter 1: Introduction

be anything from a home aquarium to salmon hatcheries that annually produce millions of fry. This book will primarily focus on RAS that are designed for the production of food fish under intensive culture conditions. A RAS for food production is a protein conversion factory; it takes the inputs of water, electricity, and feed, and outputs edible fish. The byproduct of turning feed into fish is a number of waste products that must be removed or treated in order to maintain a healthy living environment for the fish. The goal of the entire system is to treat these wastes and maintain a high-level of water quality that allows fish to be healthy under reduced water exchange conditions. The primary challenge of designing the system is to maintain this high-level of water quality while simultaneously minimizing costs.

RAS have a number of advantages over traditional aquaculture systems. They use relatively little water and space to produce a large amount fish. They can effectively treat, collect, and recycle waste streams such as fish waste. Their isolation from existing bodies of water makes it easy to prevent fish escapes and excludes predators. Similarly, the environment inside the farm is independent from the weather outside. As a result, environmental conditions of the farm can be carefully controlled and monitored, creating the opportunity to tailor the system to maximize fish growth and production. Biosecurity procedures that aim to protect the fish against disease outbreaks can also be tightly controlled, eliminating the need for vaccines and antibiotics. Finally, the systems lend themselves to high levels of automation in order to limit labor expenses, and they can be located near urban markets, limiting

Chapter 1: Introduction

transportation expenses. The downsides of RAS are that the initial infrastructure start-up costs are high, the operational costs and electricity use are relatively high, and the system must be operated by experienced staff. Nonetheless, RAS is a growing industry with businesses sprouting up all over the world. These businesses are hoping to take advantage of this well-developed technology to meet a growing worldwide demand for fresh, healthy fish.

Some species of fish are more suitable for RAS than others; the ideal species should be able to handle adverse water conditions, thrive at high stocking densities, and demand a high price on the wholesale market. Tilapia are generally considered an easy species to grow in RAS. They have a high tolerance for less than suitable water quality, and they can be sold live to grocery stores for a premium price. Other species that have been grown in RAS include rainbow trout, Atlantic salmon, yellowtail, turbot, shrimp, sole, and barramundi, among others. Research is continuing into new species of fish such as cobia, black sea bass, tuna, Coho salmon, and many others, but many challenges still wait before commercialization of these species can be fully realized. A novel fish species should be avoided, as many farms have failed based on the assumption that knowledge about the species can be obtained in tandem with it being grown in a commercial system for the first time. Instead, the water quality needs, growth rate, reproduction process, and market price for a specific species should all be as close to a known thing as possible before beginning to grow it in a commercial RAS.

Chapter 1: Introduction

Operating and building a RAS is expensive; otherwise, there would be more farms in more places. The first and biggest expense is the cost of building the entire water treatment system and all of the accessory equipment needed to feed, handle, and process the fish. On a per kilogram basis the cost of construction for RAS is higher than any other form of aquaculture, including sea cages. There are many tradeoffs to consider during this initial investment stage. For example, higher quality equipment that lasts longer and requires less maintenance also has a higher upfront cost. Also it's important to understand the economies of scale; the larger the farm the more cost effective it will be on a per kilogram of production basis. Generally speaking, a minimum production goal of 100 MT per year is needed to achieve profitability and as the production goal increases above 100 MT so does the profitability. Once the system is built the major operating expense is feed, followed by labor, electricity, fingerlings, and oxygen. Since feed is almost half of the operating budget it is important to do everything possible to maximize the fish growth for each kilogram of feed put into the system. A good farm design can also lower expenses related to labor, electricity, and oxygen.

A RAS requires a unique set of skills to start up and operate. First, there needs to be a savvy businessman/woman that understands the importance of making money, signing good contracts, and raising investment funds. Next you need a meticulous engineer who understands construction processes and equipment operations so that they can make the correct design decisions throughout the farm construction stages.

Chapter 1: Introduction

Finally, you need a day-to-day manager with a farmer's mentality, someone who is constantly observing, taking care of the small things, and planning for all eventualities. Very rarely do these three traits exist in the same person, this is why a good owner hires competent staff to fill skills gaps they may not have or they find project partners who can be responsible for core areas of the farming business where their expertise is lacking.

Over the past couple of decades many farms have invested millions of dollars into building what they advertised as state of the art systems to only fail two or three years later. The problem is many of these farms have tried to reinvent the wheel and have invested considerable money into technology research and development, or even worse built an unproven system without any research. The problem with this is that the capital requirements for the research and development are so high that the farm can never earn enough money to payback both the operational costs of the farm and the capital costs of the technology development. The good news is the wheel does not need to be reinvented and there are dozens of technology providers that can provide the necessary knowledge to plan and build a profitable RAS. This is the result of a maturing industry; technology providers are now creating a proven track record of success that they can then use to market themselves to potential future clients. This is good news since the focus for a RAS owner and operator should be on farming and not technology development. Side projects and experiments will invariably occur, but operating a profitable farm should be the primary goal.

Chapter 1: Introduction

Misallocation of capital to technology development is just one of many mistakes that most first time farmers make. Although each farm is unique there are many mistakes that are repeated over and over again. Perhaps the most common mistake is assuming an unrealistic market price for the fish. Most farmers are proud of the product that they grow, and they believe that it will be able to command a premium price due to its high quality and sustainability. However, this requires two basic assumptions, that the consumer can easily distinguish quality differences and that they care about sustainability; often times neither is true. Another issue is that prices may fluctuate seasonally or when new competitors enter the market. Assuming a high sale price for the fish can make any business plan look great, but when reality strikes, the revenue predictions will underperform and the business will not be sustainable.

Another common mistake is undercapitalization in the initial farm build out. As a result the undercapitalized farm is not large enough, doesn't adequately maintain water quality, or begins falling apart soon after start up. This either leads to farm failure in two years or it requires an additional capital infusion to fix all the problems, which then ultimately leads to a delay in production and a delay in the return on investment. Inevitability these delays end up costing more than if the farm had just been built correctly the first time. Although it can be hard to bite the bullet, sometimes it is better to invest in proper equipment and facilities the first time around. This is not so difficult if you craft an honest budget that accounts for all

Chapter 1: Introduction

expenses plus contingency cost overruns. A final common mistake is mis-selection of the initial site. Common issues include choosing a location where the water source does not have an adequate flow rate for expansion or the flow rate is so inadequate it takes 100 days to fill the entire system for the first time. Another common issue is source water quality changing seasonally or decreasing in quality when upstream neighborhoods begin polluting it.

This is just a sample of things to keep in mind while building your first farm. However, by following the proper design process steps outlined in this book many of the first timer mistakes and hiccups can be avoided. To an experienced business owner, engineer, or farmer some of the information and advice in this book may seem obvious or simply common sense. However, if it is covered in this book it is because someone at some point didn't follow that common sense advice when they should have. Be aware of and avoid the pitfalls from the past. Similarly, take the time to do the small things during the planning and budgeting stages in order to ensure that ultimately the enterprise will be a successful and profitable RAS business.

Chapter 2: RAS Design Steps

Careful planning and preparation are required to construct a successful RAS facility. Many headaches can be avoided by learning from the past failures of farms and not repeating the same mistakes. Many farms have failed by ignoring obvious shortcomings of the species they chose, the site they selected, or the design they built. Proper planning, site vetting, research, and pilot tests can avoid these pitfalls. Even in the cases of good planning new problems can arise, but contingency plans and budgets allow the farm to adapt and thrive. As an added bonus, taking the correct planning steps will provide certainty to investors and project partners, making fundraising that much easier. Through the entire planning process it is important to continuously evaluate the farm's business prospects and to progress through the planning stages in the correct order to ensure a crucial detail is not overlooked or ignored. Moving forward without first addressing a problem in the site, market, or system design is the easiest way to insure the farm will eventually fail.

It is important to remember that the initial design decisions you make for your farm do not happen in a vacuum, as one decision will affect another. At times you will have to consider tradeoffs and compromises to achieve one goal that may be in conflict with another. In these cases knowing which way to go can be difficult, you will have to rely on your team's knowledge and experience to act as your guide to make the right choice.

Chapter 2: RAS Design Steps

A good starting point to begin understanding the scale of the farm is to define a budget, annual production goal, or available footprint. A production goal is the most useful metric from which to begin the design of a farm, as it assumes that you have done research into how much product you think you can sell, and that you know what scale needs to be reached for the farm to be profitable. Generally speaking, a small farm would produce between 150 to 250 metric tons (MT) of fish per year, and a large farm would produce upwards of 1000 MT per year. Along with a production goal, a target species and site location are the first three items needed to begin crafting a RAS project scope. In large part these three components will shape the rest of design process.

Selecting a target species early in the design process will reduce the need for changes later on and it could easily have an effect on the production goals, site selection, and business plan. Each species is going to have its own acceptable water quality standards, stocking density, and growth rate, along with other species-specific eccentricities. A coldwater species like a salmon or trout requires a very different environment than a warmwater species like Tilapia, this influences the equipment selection, the space needed, and the water process flow. Similarly, a saltwater system will look very different from a freshwater system, even if both are set up for the same species. Important considerations for choosing a species are the existing aquaculture knowledge base, the current market price, and possible market competitors. These last two considerations tie in with the business plan. The selected species may also have an important influence on the farm location. For instance,

Chapter 2: RAS Design Steps

it is easier to build a farm for a warmwater, saltwater fish like Yellowtail in an area where there is easy access to warm seawater. Similarly, having a large customer base nearby where your particular species will be sold is a further asset. If possible, visit other farms already growing your target species, a few minutes looking at another farm and talking with an experienced operator can be invaluable in guiding you to the correct decision.

After determining a production goal and a target species the next step is to choose a suitable location. When selecting a farm location careful consideration should be given to water sources, water discharge capabilities, power supply, truck access, and permitting (Table 1). This crucial step is often ignored and leads to the failure of far too many farms. Often the site choice is considered an afterthought, after all, the farm is supposed to function as a self-contained, enclosed system. Unfortunately this is not really the case, the farm location needs to meet certain basic criteria or it will fail. First and foremost there must be a consistent high quality source of water. Ideally the flow rate and water quality will be consistent throughout the year. A water source can be a well, surface water, municipal water, or any other water supply. The cost per liter of water may be zero or it may be considerable if it is being obtained from certain municipal water sources. If you are in anyway dependent on someone else for your water supply, such as a city, make sure you have contracts in place to ensure that you will have a continuous, high quality supply of water. In many cases the incoming water will need to be at a minimum disinfected and in some cases treated further. This is not a

Chapter 2: RAS Design Steps

problem, but if water supply treatment is necessary it should be known beforehand and not after a site has been selected.

Next thing to consider is the ability to dispose of wastewater, all farms discharge water along with other waste streams. In some cases it may be as simple as discharging the water into an outside treatment pond or into a sewer system. It is best to understand local wastewater discharge rules and regulations to understand what is and isn't acceptable. In some cases you may need to treat the wastewater before discharging it. Just as before this is not a problem, but needs to be known beforehand. The farm should also have access to suitable power, ideally high voltage and three phase. The price of electricity may change during the time of year and even during the time of day, understanding these rate changes is important. Similarly, you may be able to receive special electrical rates as an industrial user. Truck access is important for regular deliveries of feed, and outgoing shipments of fish. This will also effect how you choose to layout the farm on the chosen site.

Finally, you will want to have a good understanding of local permitting laws related to aquaculture. In many cases local government employees will have no experience with permitting and approving aquaculture operations and there may be very little information available for what is or isn't allowed. In these cases you will want to talk to as many people as possible and see if there are any similar operations already in existence in the area. This may be the most difficult part of the site selection process to quantify, and the part where a leap of faith may be required.

Chapter 2: RAS Design Steps

Table 1. Site considerations.

Item	Details
Water source	Flow rate, complete water quality analysis ideally from multiple samples, cost of water, future changes in flow rates or water quality.
Wastewater discharge	Sewer or wetland discharge, local requirements for waste discharge.
Power supply	Three phase or single phase, available voltages, need for a transformer, cost per kilowatt-hour.
Truck Access	Truck turn around capabilities, local road limits on sizes and weights of vehicles.
Permitting	Local zoning, rules and regulations, land use limitations.
Distance to market	Average cost of transportation.
Site size	Expansion potential.

Just as important to the aforementioned steps is having a proper business plan. This book won't go into much detail on the specifics of a business plan, as there are already other excellent resources available for how to write a business plan (Appendix H). Looking at example business plans is often the easiest way to understand what one should look like, see if you can find one related to a business in the agriculture field. The basic business plan will include estimated revenues, costs, and profit for the first five years. Further information includes the marketing plans, target audience, risk assessment, and competition analysis. A rock solid business plan will be needed if you are looking for outside investments. Also, keep in mind that the business plan development happens in parallel with the farm design process and they will influence each other. For example, a final construction budget number is needed to understand the profitability of the business.

Chapter 2: RAS Design Steps

Once a species has been selected, a site has been evaluated, a production goal chosen, and a business plan formulated, the next step is to create a production plan and process flow diagrams. A production plan works backwards from the production goal to understand what the fish system needs to look like. It provides the necessary design criteria to begin sizing equipment. Inputs into the production plan are fish growth rate, feed rate, stocking density, fingerling size, harvest size, and estimated mortality. Using these numbers, the water volume needs, ammonia production, and oxygen consumption of the farm can be calculated.

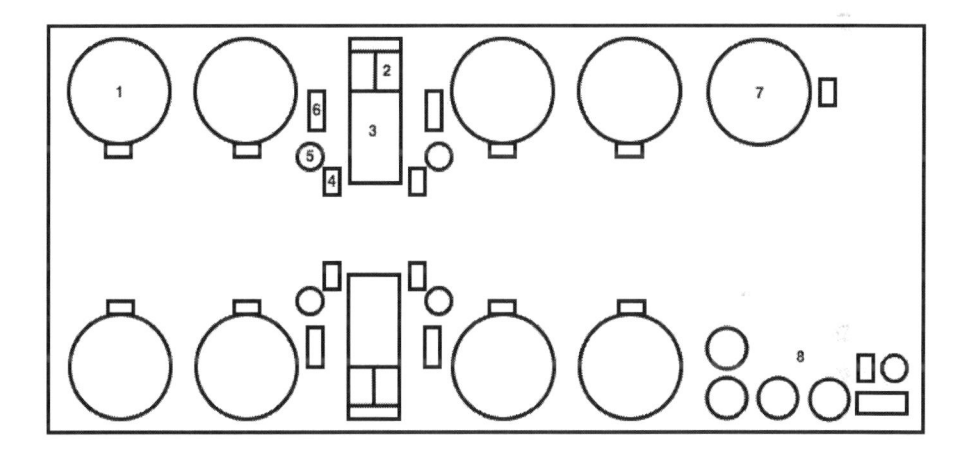

Figure 1. Example farm layout: 1.Tank, 2.Drum Filter, 3.Biofilter, 4.Pump, 5.Oxygen cone, 6.Ultraviolet Filter, 7.Purge System, 8.Nursery System

A process flow diagram looks similar to some of the flow charts made in this book, it begins to explain the path of water flow and the general specifications of the equipment that is needed. These two pieces of information will then lead to a general site layout and equipment list. A general site layout will have

Chapter 2: RAS Design Steps

dimensions of all equipment and their approximate location, it will also give an idea of the building size and what supporting equipment may be needed (Figure 1). An equipment list will include model numbers, power requirements, flow rates, and dimensions. Once an equipment list is in hand, it is time to put together a construction estimate and approximate timeline. The construction estimate can be rough at this point, but it should attempt to get within 10-15% of where the total final costs will be. A professional contractor or construction manager is best suited to estimate a construction timeline. See Table 2 below for what initial information is needed before committing to detailed engineering and construction.

Table 2. Initial planning information needed before construction.

Item	Details
Site details	See table 1.
Project goals	Species, production goal.
Process flow	Water flows, power needs.
Production plan	Fish growth rate, feed rate, time to harvest, ammonia production, oxygen, consumption, mortality estimate.
Plan view/ layout	Rough dimensional drawing of tanks and equipment.
Equipment list/ Bill of materials	Organized list of materials needed.
Budget	Construction and operational costs.
Business plan	Estimated revenue, profit, and expenses.
Construction timeline	Critical path and labor needs.

After these initial steps have been completed it is time to evaluate whether this is still a viable business proposition and if not what changes can be made to improve the business

Chapter 2: RAS Design Steps

prospective of the farm. If it is feasible, it is then time to complete detailed engineering and design. This process is iterative and will require a back and forth with owners, designers, contractors, and operators. Again, taking the time to properly complete detailed engineering can save headaches in the future. Detailed engineering includes a set of detailed drawings that are dimensionally correct and show all pipes, concrete, electrical lines, and equipment locations. The drawings should be complete enough so that any contractor could use them to correctly build the farm. The detailed engineering also includes a specification list that calls out each component needed and the quality specifications that must be met for its procurement and installation. For all of this to be completed a soil report will most likely be needed since it will determine concrete and excavation needs. Furthermore, a licensed engineer will most likely need to be engaged to check off on any structures or other pieces of the construction that could be considered liabilities. Financial institutions including banks will want the involvement of licensed engineers, electricians, and contractors to make sure the construction of the facility is completed according to local code and regulations.

After detailed engineering, the entire set of drawings and specifications should be sent to a number of contractors as a request for proposal (RFP). The RFP gives them the necessary information to accurately bid on the project. Each of these contractors will then come back with their best price on how much it will cost to build the farm. If you serve as your own general contractor then you can piecemeal out the RFP to

Chapter 2: RAS Design Steps

various subcontractors, for instance only send the electrical specifications to electrical contractors.

The detailed engineering and RFP process is how traditional construction is done. However, this does not always work for aquaculture projects for a couple of reasons. First, most contractors have no experience with building fish farms, as a result they can not offer accurate pricing and they will pad their budgets to anticipate any unforeseen obstacles. They may also not have access to good prices on some of the equipment that is needed. Second, an aquaculture farm is not always a big enough construction project to warrant going through this entire engineering process. The engineering costs may be too high relative to the total farm cost. In these cases it may be possible for the owner to design and build the system themselves, solving any engineering problems as they pop up during construction. Many farms are constructed in this way and there success is largely dependent on the aptitude and knowledge of the lead foreman who is building the farm.

After receiving proposal bids for the farm it is best to review the design again and see if there are any possible changes that would result in further savings. A good contractor will usually offer a few suggestions for where costs can be saved in the design. This is also the final step where one can decide if building the farm is still a viable investment. Once construction costs begin accruing there is no turning back without incurring some losses.

Chapter 2: RAS Design Steps

Construction time will depend on the size of the farm. As the construction process is taking place an operational plan for the farm should be put together. This should included details such as the harvest timings, equipment maintenance schedules, biosecurity protocols, fish handling procedures, and descriptions of any other critical processes. Having documentation for the operation of the farm makes training easier and allows for smoother farm operation and troubleshooting. It is possible that farm operation will occur before the farm construction is completed. Often the hatchery and nursery can begin operating as the final touches are being placed on the larger grow-out systems and the building. This decreases your time to first harvest. Once the system has been started up and as production begins a final evaluation should be completed to ensure every piece is working as planned and to identify any problems that need to be fixed before they become long-term headaches.

One step that has been omitted that others frequently include is the design, construction, and operation of a pilot system for a proof of concept. The reason this has been omitted is that if you are using a previously cultured species and have followed the above steps then there will be very few uncertainties and the risks are minimal. Universities and government labs have done a great deal of testing with numerous species in RAS. If you are planning to use one of these species along with proven RAS engineering concepts such as those described in this book then a proof of concept is not necessary. A pilot test may be needed if a new species or revolutionary system is being tested. However, this should only be attempted if you are

Chapter 2: RAS Design Steps

prepared to lose as much money as you plan to put into the pilot system. A pilot system can quickly turn into a money hole that requires years of research and development to troubleshoot. This can be fun, but is not a good way to turn a profit unless you are in it for the long haul and have an idea that ultimately turns out to be viable.

The entire RAS design process should be properly managed and documented. This may mean hiring or contracting a project manager to help with the process. If it is to be done internally there are a few concepts to keep in mind. The first priority is to manage the scope of the project, this starts during the initial planning stage and continues throughout the entire project. The scope defines what is to be accomplished, in what time frame it will be accomplished, and how much it will cost. Any change to the scope should also warrant a change to the timeline, budget, and/or resources needed. Timelines help with managing critical path items, task lengths, proper order of tasks, and scheduling. Budget management includes managing costs, both estimated and actual, and contingencies. Resources include people, equipment, and construction materials. Making the effort to define the scope and properly manage the design and construction process is well worth the time spent. A huge number of changes occur between project inception and final start up. Being able to track and understand these changes will lead to the execution of a cheaper, faster, better RAS farm.

A few final notes: during this entire process you should pay careful attention to any contracts that you need to either write yourself or agree to as a signee. Pay careful attention to the

Chapter 2: RAS Design Steps

language in the contract and feel free to negotiate points that you disagree with. Limit your own liability when possible, and make sure project partners are held responsible for mistakes, cost overruns, or delays. Unfortunately a single bad contract and a bad partner can bankrupt a farm. If this whole process seems overwhelming consider working alongside an experienced aquaculture engineer or equipment provider. Take your time, sweat the details, and stick to the plan, your RAS farm is worth the time and effort.

Chapter 3: Production Plan

The production plan outlines the design criteria that the farm must meet in order to accomplish its production goals. The plan works backwards from the production goal to calculate critical values such as ammonia production, oxygen consumption, carbon dioxide production, and total system water volume.

In order to calculate these values a number of biological parameters for your chosen species must first be known (Table 3). Parameters include maximum stocking density, feed conversion ratio (FCR), daily feed rate, protein content of feed, estimated growth rate, initial stocking weight, and final harvest weight. The amount of feed that enters the farm each day is directly proportional to ammonia production, carbon dioxide production, and oxygen consumption. The feed rate also dictates the growth rate. Once you know the growth rate of the fish you can then calculate the total carrying capacity that the farm must support in kilograms of fish.

Chapter 3: Production Plan

Table 3. Productions plan inputs.

Biological Parameters	Normal values
Stocking density	30 – 120 kg/m^3
Stocking size	Egg – 50 g
Final size	500 g – 3 kg
Daily feed rate	1 – 10% body weight/day
FCR	1.0 – 1.5 kg$_{feed}$/ kg$_{fish}$
Protein content of feed	35 – 50 %
Net protein utilization rate	50 – 65 %
Biofilter nitrification rate	0.2 – 0.5 g$_{TAN}$/m^2/day
Oxygen consumption rate	0.2 – 0.6 kg$_{O2}$/kg$_{feed}$
Water exchange rate	1 – 10 % / day
System hydraulic retention time	30 – 60 minutes

These details provide the necessary information to specify the number and size of tanks, the water flow rates, the size of the biofilter, and the size of the oxygenation system. All other design specifications follow from these calculations, such as the size of UV filters, the size of the drum filters, the daily water and energy needs of the farm, and all other essential components. In addition to allowing for the specification of equipment, the production plan influences operational procedures such as how often fish are handled, how many kilograms of fish are purged per week, and most importantly how much feed is used per day. A production plan may need to be formulated for multiple life stages of the fish as it moves throughout the farm. For example, there may need to be a hatchery production plan, and a grow-out production plan. This is most appropriate when the biological needs of the fingerling fish are drastically different than mature fish; fingerlings often require lower stocking densities, and higher daily feeding rates.

Chapter 3: Production Plan

It is also important to understand that the production plan may have to change after startup. The assumptions for the input biological parameters may not hold true during actual operations, the FCR may be slightly less or the stocking density may not be as high. The adjustments could affect the annual production of the farm for better or worse. Creating a good production plan requires a lot of knowledge and experience. Since this information forms the backbone of your farm's design it is suggested you work with a well-recommended RAS designer during this step of the process. Below is an example set of calculations that you can use as a guide for your production plan.

Table 4. Example production plan inputs. Values are estimated from past research or are known constants (each species will have slightly different values, unknown values may need to be inferred).

	Variable	Value	Units
Annual production goal	G	1000	MT_{fish}
Stocking density	d	100	kg_{fish}/m^3
Daily feed rate	FR_d	0.02	$kg_{feed}/kg_{fish}/day$
Feed conversion rate	FCR	1.2	kg_{feed}/kg_{fish}
Daily water exchange	Ex	5	%
Hydraulic retention time of system	HRT	30	Minutes
Protein content of feed	P_{feed}	0.50	$kg_{protein}/kg_{feed}$
Fish protein utilization rate	U_{feed}	50	%
Nitrogen ratio constant	N	0.16	$kg_N/kg_{protein}$
Biofilter nitrification rate	n	0.5	$g/m^2/day$
Oxygen consumption rate	$O2C$	0.6	kg_{O2}/kg_{feed}
Carbon dioxide ratio constant	$CO2P$	1.4	kg_{CO2}/kg_{O2}

Chapter 3: Production Plan

System Volume Calculation

$$Daily\ production = \frac{G}{365\ days} = \frac{1000\ MT_{fish}}{365\ days} = 2.74\ \frac{MT_{fish}}{day}$$

$$Daily\ feed = Daily\ production \times FCR = 2.74\frac{MT_{fish}}{day} \times 1.2\frac{kg_{feed}}{kg_{fish}}$$

$$= 3.29\ \frac{MT_{feed}}{day}$$

$$System\ biomass = \frac{Daily\ feed}{FR_d} = \frac{3.29\ \frac{MT_{feed}}{day}}{0.02\ \frac{kg_{feed}}{kg_{fish}}\Big/day} = 165\ MT_{fish}$$

$$System\ volume = \frac{System\ biomass}{d} = \frac{165\ MT_{fish}}{100\ \frac{kg_{fish}}{m^3}} = 1650 m^3$$

$$Daily\ water\ exchange = System\ volume \times Ex = 1650 m^3 \times 5\frac{\%}{day}$$

$$= 82.5\ \frac{m^3}{day}$$

$$Water\ circulation\ rate = \frac{System\ volume}{HRT} = \frac{1650\ m^3}{30\ min} = 917\ \frac{liters}{second}$$

Chapter 3: Production Plan

Ammonia Production Calculation

$Daily\ total\ ammonia\ nitrogen\ production$
$$= Daily\ feed \times P_{feed} \times (1 - U_{feed}) \times N$$

$$= 3.29 \frac{MT_{feed}}{day} \times 0.5 \frac{kg_{protein}}{kg_{feed}} \times 50\% \times 0.16 \frac{kg_{TAN}}{kg_{protein}} = 132 \frac{kg_{TAN}}{day}$$

$Biofilter\ media\ surface\ area = Daily\ TAN\ production \times n$
$$= 132 \frac{kg_{TAN}}{day} \times 0.5 \frac{\frac{g_{TAN}}{m^2}}{day}$$

$$= 65,800 m^2$$

Oxygen Consumption & Carbon Dioxide Production Calculation

$Daily\ oxygen\ consumption = Daily\ feed \times O2C$
$$= 3.29 \frac{MT_{feed}}{day} \times 0.6 \frac{kg_{O2}}{kg_{feed}} = 1970 \frac{kg_{O2}}{day}$$

$Daily\ carbon\ dioxide\ production$
$$= Daily\ oxygen\ consumption \times CO2P$$

$$= 1970 \frac{kg_{O2}}{day} \times 1.4 \frac{kg_{CO2}}{kg_{O2}} = 2758 \frac{kg_{CO2}}{day}$$

Chapter 4: Water Quality

The goal of any RAS is to provide adequate water quality. Good water quality means healthy fish and healthy fish grow faster. Fast-growing fish are more profitable allowing for the system to produce more revenue in less time. Each step of the water treatment process has a specific role in maintaining certain water quality parameters. The battle to maintain water quality is never-ending. As stocking density increases it becomes more difficult to manage water quality, and there is less room for error.

Some water quality parameters change slowly over the course of days or weeks like nitrates, while others like dissolved oxygen can change drastically in minutes. Each parameter comes with its own concerns and considerations. Many parameters are interlinked, for example, pH, alkalinity, carbon dioxide (CO_2), and ammonia (NH_3) levels will all change if any one of the four changes. Properly managing water quality is very difficult and even experienced operators will find themselves scratching their heads as to why one parameter is out of the desired range. The key is to pay attention to details and catch changes in water quality early.

Chapter 4: Water Quality

Table 5. Typical Water Quality Values.

Parameter	Value
Temperature	16 - 28 °C (depends on species)
Salinity	0 - 35 ppt (depends on species)
Dissolved Oxygen	> 5 mg/L or > 60%
Dissolved Carbon dioxide	< 15 mg/L
Total gas pressure	< 103%
TAN	0 – 2.5 mg/L (depends on pH and species)
Nitrite	< 0.5 mg/L
Nitrate	0 - 150 mg/L
pH	6.8 – 8.0
Alkalinity	50 - 200 mg/L
Hardness	100 - 300 mg/L
Total dissolved solids	0 - 400 mg/L
Total suspended solids	0 - 80 mg/L (depends on species)

Above is a list of typical water quality values for the most commonly measured parameters (Table 5). These parameters can be affordably monitored with sensors and water testing kits. Additional parameters can be evaluated by sending samples to a water-testing laboratory on an as needed basis. All of these values are general guidelines; all species will have different preferences and tolerances for each of these values. Furthermore, most species will be able to tolerate values outside of the given ranges for short periods of time, but to maximize fish health one should stay within these ranges.

Temperature and salinity are both easily measured on a continuous basis with electronic sensors. Although most species can tolerate a range of temperature and salinities each one has an optimal target range that will maximize growth.

Chapter 4: Water Quality

Temperature is a physical quantity that approximates the average kinetic energy of molecules, it is typically measured in either degrees Fahrenheit (°F) or Celsius (°C). Salinity is the measure of the mass of salts dissolved in the water, most commonly sodium and chloride. Salinity is typically measured in parts per thousand (ppt), percent (%), grams of salt per kg of water (g/kg) or sometimes as conductivity (S/m).

Dissolved oxygen (DO), dissolved carbon dioxide (pCO2), and total gas pressure (TGP) can be measured with different levels of ease. Oxygen can be continuously monitored with a range of electronic sensors, typical units of measurement include milligrams of oxygen per liter of water (mg/L), parts per million (ppm) or percent of saturation (%). The saturation concentration of oxygen is dependent on the water's temperature and salinity, lower temperature and lower salinity increases the saturation concentration. Dissolved oxygen in equilibrium with the environment is usually around 8-10 mg/L. Dissolved oxygen will go up and down throughout the day and night, mostly falling right after feeding times and slowly increasing afterwards.

Dissolved carbon dioxide can also be measured with electronic sensors typically in milligrams of carbon dioxide per liter of water (mg/L), parts per million (ppm), or partial pressure (mmHg), but the technology is not as developed as oxygen sensors and it is not generally continuously monitored. Carbon dioxide when dissolved in water becomes carbonic acid; because of this the concentration of carbon dioxide in the water is closely related to the pH and alkalinity of the system.

Chapter 4: Water Quality

As a result carbon dioxide levels can be independently tracked by also monitoring the pH and alkalinity of the system. Carbon dioxide typically peaks after feeding events and is inversely related to oxygen. Carbon dioxide is around 1 mg/L when in equilibrium with the environment, but is much higher in aquaculture systems where there is constant fish respiration occurring. New standards are still being developed to understand the acceptable level of carbon dioxide in the water; currently 10-30 mg/L is considered the upper threshold, dependent on the species.

Total gas pressure measures the partial pressure of gases dissolved in the water, these gases are primarily nitrogen, oxygen, and carbon dioxide. Total gas pressure can be easily measured with an electronic sensor and is often checked only intermittently, it is typically measured as percent saturation (%). High levels of nitrogen gas, which can be indirectly estimated by total gas pressure, are problematic for fish and causes various disorders. As a result total gas pressure should be kept below 103% and typically it only exceeds this if there is air being injected under pressure somewhere in the system.

Total ammonia nitrogen (TAN), nitrite (NO_2) and nitrate (NO_3) are all present in RAS systems and each has a different level of toxicity for fish. Typically the concentrations of all three are measured in mg/L or ppm. All three can be measured by chemical water testing, after following a set of procedures and mixing a water sample with premeasured quantities of chemical drops or powder, the water color of the sample will then change. The color can then be compared to a color wheel or

Chapter 4: Water Quality

measured in a spectrophotometer to gauge how much TAN, NO_2, or NO_3 was present. New sensors can electronically monitor TAN concentrations in freshwater systems, but daily water sampling and testing of each by color analysis is still the standard procedure.

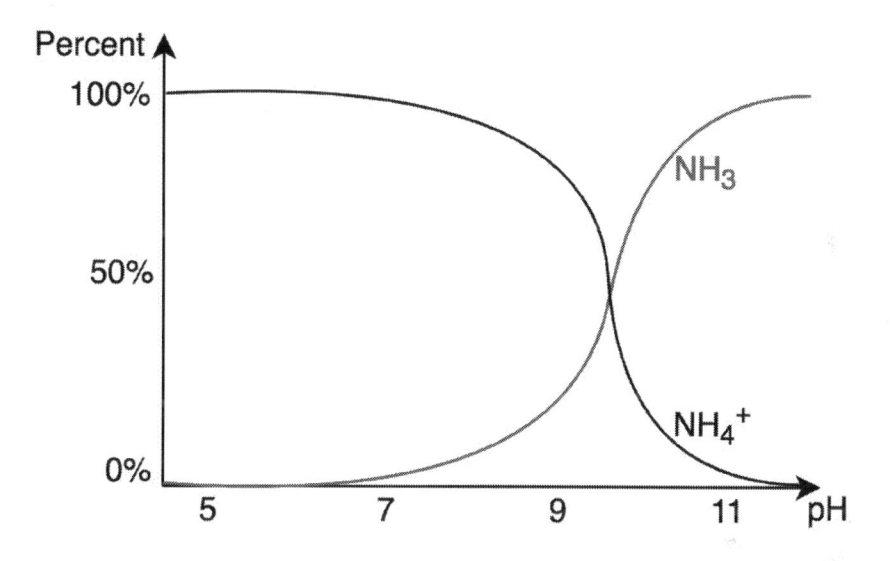

Figure 2. Approximate concentrations of ammonium (NH4) and ammonia (NH3) as a function of water pH.

TAN is composed of both ammonia (NH_3) and ammonium (NH_4), both forms are present in all aquaculture systems, the relative concentrations of each are dependent upon the systems pH (Figure 2). TAN is excreted by fish as a waste product and the un-ionized ammonia form is toxic to fish at relatively low concentrations, the biofilter performs the work of converting TAN into nitrite then nitrate to reduce its toxicity. Monitoring of the relative concentrations of each is particularly

Chapter 4: Water Quality

important during biofilter start up. Nitrate concentrations will slowly build up in the system until a steady state concentration is reached; at which point as much nitrate is being flushed out of the system each day in replacement water as is being produced by the biofilter.

pH, alkalinity, and hardness are all loosely related in aquaculture systems and the relationships between them involves a number of chemical reactions which are often confusing to understand. pH is the measure of hydrogen ions (H^+) present in water, it is measured on a logarithmic scale, a change of one unit is equal to a ten fold change in the concentration of hydrogen ions. A pH of 7 is considered neutral, higher pHs are basic and lower pHs are acidic. Aquaculture systems often are maintained somewhere between a pH of 6.5 and 8.0, also known as the zone of life. pH can be easily monitored with electronic sensors or colorimetric indicator test strips. pH should be regularly monitored as it is both affected by and affects a number of different water treatment processes such as biofiltration and carbon dioxide stripping.

Alkalinity is a measure of the ability of the water to neutralize an acid, also known as the buffering capacity. In aquaculture systems this is typically composed primarily of the carbonate carbon system and it can be approximated with the below equation.

$$Alkalinity = [HCO_3^-] + 2[CO_3^{2-}] + [OH^-] - [H^+]$$

Chapter 4: Water Quality

A high alkalinity will neutralize acids and allow the pH to remain stable even if acids are entering the system. Alkalinity is measured by taking a water sample, adding a pH indicator dye, and then performing a titration. As acid is added to the sample the alkalinity will buffer the solution until the pH changes dramatically and the dye changes color. By knowing exactly how much acid was added to the sample we can calculate the alkalinity from this test. Alkalinity is typically measured in either milligrams per liter of calcium carbonate (mg/L $CaCO_3$) or milliequivalent per liter (mEq/L).

The interaction between alkalinity and pH is a good example of the kind of chemical interactions that occur in any solution. For example, as hydrogen ions are added to a solution a new equilibrium point is reached between a variety of different molecules that are in that same solution. The concentrations of ammonia and ammonium will change (Figure 2), as will the relative concentrations of carbonic acid ($H_2CO_3^*$), carbonate (HCO_3^-), and bicarbonate (CO_3^{-2}) (Figure 7), and so will the concentrations of a number of other molecules. These kinds of interactions and changes in equilibrium points are not just limited to pH, the addition and subtraction of all ions, compounds, and molecules in solution has an affect on the other ions, compounds, and molecules in that same solution.

Alkalinity and hardness are often confused since they are closely tied together and measured with the same units of mg/L $CaCO_3$. Hardness though, is the measure of divalent ions in solution, most frequently calcium (Ca^{2+}) and magnesium (Mg^{2+}). Hardness is often tied to alkalinity because these

Chapter 4: Water Quality

divalent ions enter the system as either calcium carbonate or magnesium carbonate, which affects the alkalinity. The hardness of the water partially explains the mineral content of the water. Harder waters have high conductivity and less metal availability, and very hard waters can lead to scaling and mineral deposition on equipment and pipes. Water softener systems are used to lower hardness, and many use either ion exchange or energy intensive methods such as reverse osmosis. This is not done as a regular part of RAS water treatment, but it may be done to treat a very hard (>500 mg/L) water source before it enters the farm. Hardness can be measured irregularly with a chemical water testing kit, it can also be indirectly measured by an electronic conductivity sensor.

Solids come in variety of different sizes and can be measured in number of different ways, the smallest solids (<2 μm) are known as total dissolved solids (TDS). TDS is the measure of the total amount of minerals, salts, metals, nitrates, and dissolved organic substances that are in colloidal suspension in the water. These are the solids that cannot be filtered out by any kind of physical screen. The easiest method to estimate TDS is to use an electrical conductivity sensor, which measures the number of ions in the water giving an approximation of within 10% of the actual TDS. Units of TDS are typically reported in either milligrams per liter (mg/L) or parts per million (ppm).

Total suspended solids (TSS) is a measurement of all settleable solids. These are solids that are large enough that they will settle out of the water column with enough time under still

Chapter 4: Water Quality

conditions. In aquaculture systems TSS is mostly composed of fecal matter and leftover feed. It is measured by taking a water sample of a known volume and passing it through a fine paper mesh filter. The solids that remain on the filter paper are weighed on a dry weight basis and the solids are reported in milligrams of solids per liter of water (mg/L).

There can also be some confusing nomenclature around the names of the different types of solids and some gray area on the differences between the different types of measurements. The terms total settleable solids and total suspended solids may be used interchangeably, but each is measured in a different way. For example, in biofloc systems, where suspended bacteria colonies are used to remove ammonia, the solids concentrations are extremely high, and the amount of total settable solids is typically measured in milliliters of solids per liter of water (ml/L). A one-liter sample of water is placed in an Imhoff cone and over the course of a couple hours the solids settle to the bottom and give a volume measurement in milliliters. This is not comparable to TSS, which is measured on a dry weight basis in mg/L. For normal aquaculture solids the best approximation to convert between these two is 1 ml/L of total settable solids is equal to 13 mg/L of TSS.

How to control of all these water parameters will be discussed further in the next section. Each of the water treatment steps is aimed at directly or indirectly regulating these parameters so that they stay within the optimal zone for fish growth.

Chapter 5: Water Treatment Components

In this section we will explore the essential water treatment steps required to maintain water quality in recirculating aquaculture systems. Most water treatment steps occur along a single path known as the main flow or main treatment flow, and these steps should be performed in the correct order to ensure maximum system efficiency (Figure 3). Some water treatment components, such as heating and cooling, are not sequential or necessarily a part of the main treatment flow and these processes are known as side streams. The following sections will approximate the path the water takes starting at the tank drain and ending with the tank inlet. Along this path the water undergoes five key processes: solids removal, biofiltration, degasification, oxygenation, and circulation. In addition to these five there are multiple optional main flow and side stream treatment processes that the water may also undergo. Ideally all steps should include a certain level of redundancy in their design so that if any one piece of equipment fails the entire system can still remain operational.

Chapter 5: Water Treatment Components

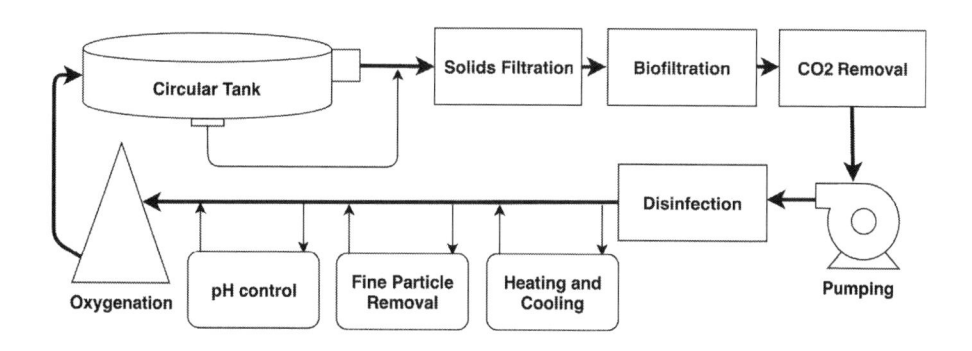

Figure 3. RAS water treatment flow chart.

Tanks

Tank selection requires two important components, dimensions, and material choice. Additionally you will need to consider how many individual tanks and what sizes will be required for your chosen species production plan. Tanks sizes can range anywhere from a 20 liter cylinder in a hatchery to a 600,000 liter, 20 meter diameter tank that require a dinghy and SCUBA team to service and clean. Generally speaking the larger the tank the lower the cost is on a per-liter basis. Most RAS tanks are either round, octagonal or somewhere in between. Because of their rare use in RAS this book will not discuss raceway or mixed-cell raceway tanks.

The relative dimensions of a tank are crucial to allow for self-cleaning of solids. Self-cleaning significantly reduces labor costs, while improving water quality. For this reason round tanks with a circular water flow are most commonly used, tanks with eight sides or more can be considered round for all intensive purposes. By setting up a circular flow in a round tank

Chapter 5: Water Treatment Components

the centrifugal forces of the current will carry the heavier-than-water solids towards the center of the tank where they are then removed via a center drain. In order to have optimal water flow the depth of tank should be between one-fifth and one-half of the tank's diameter. Both physical testing and computational fluid dynamic tests have shown that these ratios allow for the best hydraulics to facilitate self-cleaning. Having a slight five-degree slope on the tank floor towards the center drain also helps facilitate the cleaning of the tank and makes completely draining the tank easier when cleaning or maintenance needs to occur.

In addition to the correct tank dimensions and shape you also need an inlet and outlet system that facilitates the creation of a circular flow. The Cornell dual drain is the current standard for this in tanks that are larger than 12 m^3 and 4 meters in diameter. In the Cornell dual drain design 10 to 50 percent of the water drains from a center floor drain and the remaining water drains into a side box (Figure 4). The side box is literally a box on the side of the tank that drains water from the surface. The advantage to this system is it concentrates a majority of the solids into the center drain flow where it can then be easily filtered out.

The reason that all of the flow cannot to be drained out of the center drain is that when the water flow rates are at an exchange rate of one complete tank turnover per hour or less the rate of draining is so high that it will set up a powerful whirlpool in the center of the tank, and at a certain point the fish can no longer swim against the intense water flow and they

Chapter 5: Water Treatment Components

may become pinned against the center drain. The goal is to aim for a rotational velocity of approximately one meter per second (m/s). In order to achieve this velocity water should enter the tank at multiple points and the water flow should be directed tangentially to the tank wall to create a circular flow. Directional nozzles are common for this, as the flow may need to be tuned periodically to achieve even water distribution and an ideal rotational velocity.

DIA = 2h - 5h

h

Circular Tank

Sidebox
(50-90% of flow)

Center Drain
(10-50% of flow)

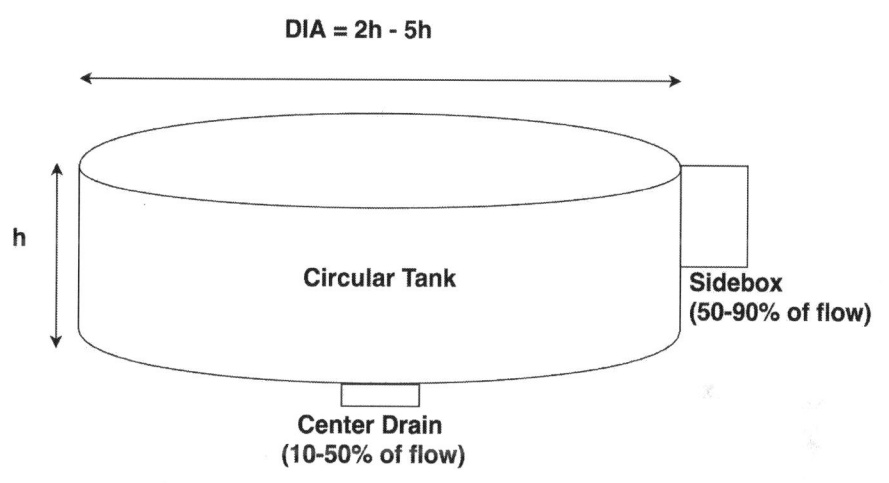

Figure 4. Tank relative dimensions and flow rates.

Tank material is an important choice that is dependent on tank size, water corrosiveness, and cost. Generally speaking there are four choices for tank materials: plastic, fiberglass, steel (glass-coated or galvanized), and concrete. Under normal circumstances all of these tank materials offer adequate corrosion resistant even in saltwater conditions, except perhaps for galvanized steel which is only appropriate for freshwater applications. A rotomolded polyethylene tank is most

Chapter 5: Water Treatment Components

appropriate when the tank has a volume below 12,000 liters and is four meters or less in diameter, at this size or smaller, plastic tanks are very cost effective. The drawback is that you will have to select from a manufacturer's line of tanks and will not be able to choose custom dimensions, but generally this is not an issue. In the four meter to seven meter diameter range both fiberglass and glass coated steel (GCS) are good material choices. In both cases the tank will be shipped to the farm in panels from the manufacturer and assembled on site.

Above seven meters in diameter, concrete and glass coated steel are the two best choices. The cost competitiveness of each will depend largely on local concrete costs. The downside of concrete is that it can be difficult to work with and requires a long cure time. Customizing inlet and outlets is more difficult than GCS and the tanks cannot be moved and reassembled once completed. GCS tanks can be expensive and require a concrete base, but they also have a long life and can be recycled if no longer needed. At tank diameters of 15 meters or greater, concrete may become more cost competitive. When building extremely large tanks proper engineering is required and experienced contractors should be used for the installation. A final option for the home hobbyist is a plastic tarp lined tank. Thick, flexible, UV protected, tarp can be purchased from a number of vendors and installed inside of a hole, cinderblock ring, or wood structure. This is not recommended for commercial growers due to the short lifespan of this material, the difficulty in obtaining a smooth tank bottom and sides, and the inability to achieve large enough single tank volumes.

Chapter 5: Water Treatment Components

The placement of tanks should be carefully considered before installation. Ideally the tank should be easily accessible on all sides with an outside working wall height of around one meter. This can be achieved by either burying the tanks below grade, or by creating a raised walkway around the tanks. Typically excavating below grade is the less expensive and better-suited solution. Smaller tanks may need to be supported on either cinder blocks or a raised bed of gravel, both work well for plastic and fiberglass tanks when routing piping below grade is not an option or is inconvenient. Glass-coated steel and concrete tanks require a concrete flooring in the tank and as a result it is best to set the center drain at grade and have center drain piping go below grade.

The sizing of tanks is more art than science at times. First the total system volume needs to be calculated during the production plan. The system volume then needs to be divided between systems and tanks. If you have too many systems it may be difficult to manage and there is more equipment to purchase and maintain. On the other hand, if you have too few systems your biosecurity is reduced and you will not have enough system redundancy in the case of equipment failure. Because of this it is difficult to say what the exact right number of tanks should be, but there are some general rules to follow. First, a system should have between four to ten tanks, and a farm is typically made up of three to ten systems. This provides enough redundancy with the ability to easily manage all of the systems. The depth and circumference of your tanks will

Chapter 5: Water Treatment Components

determine how many there are and it will influence what tank material is most appropriate.

Take the time to carefully consider tank material, dimensions, and placement. Additional considerations include the design and placement of inlets, drains, and sideboxes. The right tank will cost less, last longer, and make farm management easier.

Solids Filtration

The first filtration step after the water leaves the tank is the removal of solids. Solids are a nice way of saying fish poop or fecal matter. Generally, solids have a density slightly greater than water, are brown, and can be found in a variety of sizes from 10 micrometers (μm) in diameter to 3 centimeters (cm) in diameter. A key fact to remember is that solids are always easier to remove when they are larger, as not only can they be more easily screened, but they also sink faster. As a result a good designer should do everything they can to ensure solids do not get broken up before being filtered out. This means not going through a pump or cascading over a weir.

Solids need to be removed since bacteria will start breaking them down as soon as they leave the fish. These bacteria will consume oxygen and produce ammonia as they decompose the protein, fats, and carbohydrates in the solids, which will put an extra burden on the biofiltration and oxygenation systems. Even without the presence of bacteria, solids will begin to dissolve and fall apart, making removal more difficult and leaching nitrogen and phosphorous into the water. Using round

Chapter 5: Water Treatment Components

tanks and self-cleaning drain systems as described in the last section goes a long ways towards eliminating acquiescent zones where solids can pile up. It is likely that solids will still pile up in some corner of the system and these areas should be regularly cleaned out when found. The biofilter will also produce some solids due to the biofilm sloughing off the media, the biofilm is composed of mostly dead bacteria cells. Typically, the solids generated by the biofilter are insignificant and will get pumped back to the tanks, and then get either eaten by the fish or removed by the solids filter. Efficiently removing all solids will go a long way towards keeping the rest of the system running smoothly.

Typically the solids filtration system should aim to keep total suspended solids (TSS) between 10 to 25 mg/L. There are countless creative solutions to remove solids from the system, some methods require costly equipment that is extremely efficient, other methods can be cheap to build, but less effective or higher in labor needs. Possible solids filtration methods include radial flow separator, bead filter, sand filter, drum filter, parabolic screen, cartridge filter, or a settling basin. Each method has its advantages, disadvantages and appropriate application. All methods can be categorized into either a gravity method or a filter method.

Commonly used gravity methods for removing solids include radial flow separators and settling basins. In gravity methods, the solids settle to the bottom of a vessel and the concentrated solids are periodically removed or flushed out. In a typical settling basin, water enters one end of a rectangular vessel and

Chapter 5: Water Treatment Components

flows out over a weir on the other end. The larger the vessel and the less water flow the greater the chances of removing all sizes of solids. Larger solids will sink faster, smaller ones will sink more slowly. Reducing turbulence in a settling basin helps all sizes of solids to settle more quickly. The length of the basin does not determine its efficiency, what does matter is the cross sectional surface area, which is equal to the width of the basin, multiplied by the height. Maximizing the width and height of the basin will maximize the amount of solids that are removed. However, the major drawback of settling basins is the area required to operate them. In a commercial scale RAS the water flow rates are so high that the settling basins must be equal to the size of the rest of the farm in order to be large enough to remove most of the solids. Since this is not practical, settling basins are only used on small-scale systems. The advantage of a settling basin is that it has no moving parts and is easy to construct.

A radial flow separator (RFS) is another type of gravity solids settler that uses clever geometry to make removal more effective. The water enters the RFS going downwards towards the cone shaped bottom (Figure 5). The stilling well in the center and the overflow weir around the entire edge of the RFS make it so solids are more likely to settle and not upwell out. The larger the diameter of the unit, the more effective it is for a given flow rate. You will often see a RFS used to treat the flow of water that comes from the center drain of the tank, since this is a smaller flow with a higher concentration of solids. However, this is often only good for settling out the largest, fastest settling solids. In cases where smaller solids stills need to be

Chapter 5: Water Treatment Components

removed the water will need to undergo further solids filtration. The advantage of a two-step solid filtration process with a RFS is that it quickly removes large solids, putting less of a burden on the next filtration step. Additionally, RFS are easy to build and require no electricity or moving parts. The solids that accumulate in the bottom of the cone are regularly flushed out by manually or automatically opening the valve on the bottom.

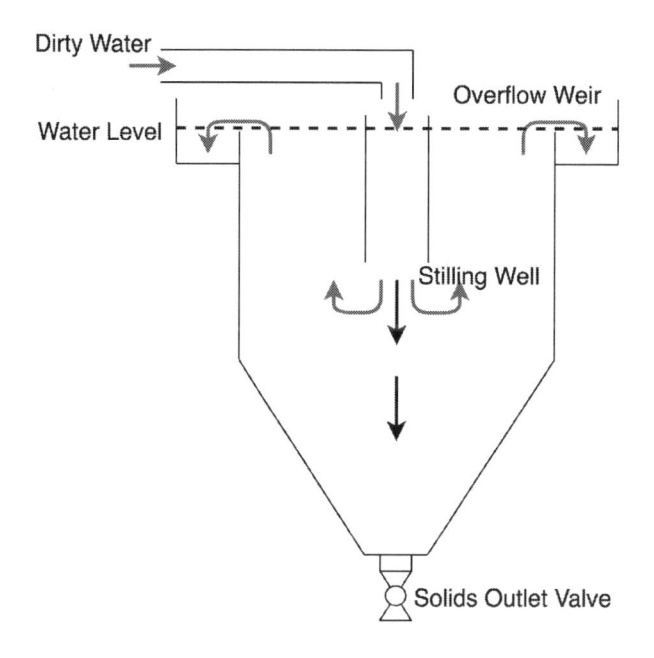

Figure 5. Radial flow separator cross-section. Gray arrows show the path of water.

The other category of solids removal devices is filters. The simplest forms of filtration are parabolic filters, which use a mesh screen to remove solids, or cartridge filters, which use a cloth bag to filter out solids. Both methods are easy to implement, but they are size limited to around 1200 liters per

Chapter 5: Water Treatment Components

minute (LPM), as above this flow rate they become impractical. Also in the case of the cartridge filter, regular backwashing or filter bag replacement is required to keep the filter from clogging. However, both of these methods are a cost effective way to filter water on smaller hatchery or hobbyist systems.

The next method of filtering solids is the use of sand or bead filters. These two filters behave very similarly to each other. In both cases, water is pumped into a round, pressurized vessel full of beads or sand. As the water passes through the media the solids are filtered out as they catch in notches and crevices between the sand/beads. As solids accumulate the pressure inside the vessel rises, the solids block the water's path as it passes through the media. Eventually the unit needs to be backwashed to remove the solids. Either shutting off the pump or rerouting the flow around the filter makes this possible. The valves on the backwash outlets are then manually or automatically switched so that the water inside of the vessel drains out to a waste line taking all of the accumulated solids with it. Frequently there is also a way to agitate the media with either air or water to further dislodge any solids that may be stuck on or between the sand/beads. Once backwash is complete the filter can then be switched back to normal operation mode.

Both sand and bead filters are commonly used as swimming pool filters, and as a result many designs are mass-produced and inexpensive. There are also aquaculture-specific bead filters that are more robust and use clever patented methods for backwashing the media. In general, bead filters and not

Chapter 5: Water Treatment Components

sand filters are recommended for aquaculture. A sand filter will remove smaller particles, but with the large amount of solids in most aquaculture systems sand filters will require backwashing too frequently to be practical. A bead filter is practical for flow rates from 200-4000 LPM, although models are available for flow rates from 40-8000 LPM. The downside of a bead filter is the need to pump the water into the unit, which breaks apart the solids, and the head loss across the unit, which requires extra pumping power. It is recommended to automate the backwashing with actuated valves that will start the backwashing process once a certain pressure inside the unit is reached.

The drum filter is the most widely used solids filter in commercial aquaculture systems. Using multiple drum filters placed side-by-side makes it possible to filter thousands of liters of water without pumping and with very little head loss. Water can flow directly from the tanks into a drum filter sump and the solids will be quickly removed before undergoing any other treatment steps. A drum filter works by having water flow in one end of a cylindrical drum and then exit by flowing over fine mesh screens that are arranged on the outside of the cylinder. Once the screen begins to clog, the entire cylinder rotates and high-pressure nozzles activate to wash the solids off of the screens and into troughs that carry the solids away to a waste pipe. In this way the screens can be continuously cleaned. The downsides to a drum filter are that they require electricity, have moving parts, and are expensive. They require the use of a high-pressure pump that is capable of delivering up to 10 bar of pressure to operate the spray nozzles that

Chapter 5: Water Treatment Components

remove the solids from the screen. Drum filters require automated controls so they backwash whenever the water level begins to rise behind the drum filter, indicating that the screens are clogged. The most commonly used screen size is 60 μm, although smaller screens may be used on hatchery systems. In general, drum filters are not economical for flow rates below 1200 LPM, however they are the only real option for flow rates over 4000 LPM. A number of manufactures specialize in making drum filters, and the larger units can cost as much as a new luxury car, but they get the job done and last a long time with regular maintenance.

Drum filters come in a variety of models, and some include a tank below the drum, while others simply rest inside of a sump. All share the same basic mechanical components, though the drive method may be a cogwheel, chain, or belt. Each has its drawbacks and benefits, but most likely the cost will be the determining factor in which one should be chosen, as they all accomplish the same task of rotating the drum. Regular maintenance includes checking the drive mechanism and the spray nozzles. Similarly the drum filter pressure pump needs to be checked regularly, as will all other pumps on the farm.

All solids filtration methods concentrate the solids they collect into what is essentially a thick slurry of solids and water. This slurry can be of significant volume in larger farms and it needs to be disposed of properly. For all methods of solids filtration, careful consideration should be given as to where solids go once they are removed via backwashing or some other method. It is best to have them flow via gravity to a single tank,

Chapter 5: Water Treatment Components

and from that tank they can be further treated which is discussed in later sections.

A single farm will have a variety of solids filtrations methods to suit the individual needs of each system. Methods vary in their efficiency, size, energy use, cost, and location in the treatment process. Take the time to plan which method will work best for your farm.

Biofiltration

Biofiltration is one of the most researched aquaculture subjects and yet perhaps still one of the least understood. The primary challenge in understanding how a biofilter operates is the ever-changing dynamics of the bacterial population that does all of the heavy lifting in a biofilter. The bacteria create a biofilm on the surface area of the biofilter, and this biofilm is composed of a community of bacteria species. These bacterial populations respond to the slightest changes in their environment, such as swings in pH, temperature, TAN concentration, salinity, and oxygen levels. Understanding these changes is crucial to operating a RAS.

At its simplest a biofilter is a vessel with a high amount of physical surface area that supports a biofilm. The bacteria in the biofilm are responsible for converting total ammonia nitrogen (TAN) into nitrite (NO_2) and then nitrate (NO_3). The purpose of converting TAN into nitrate is the relative toxicity of each substance. TAN is toxic to fish in concentrations as low as 0.5 mg/L whereas nitrate can be tolerated by fish in concentrations

Chapter 5: Water Treatment Components

up to 300 mg/L. TAN is excreted by fish in their urine, so just as solids filtration treats fish fecal matter, the biofilter treats fish urine. The bacteria convert TAN into nitrate in a two-step process, first by transforming TAN into nitrite and then nitrite into nitrate.

$$NH_4^+ + 1.5O_2 \rightarrow NO_2^- + 2H^+ + H_2O$$

$$NO_2^- + 0.5O_2 \rightarrow NO_3^-$$

The above process is generally attributed to two bacteria geneses, nitrosomonas for the first step and nitrobacter for the second step. Both bacteria are chemoautotrophs: They use nitrogen as an energy source and utilize inorganic carbon in constructing their cells. The above chemical equations also show that both steps require oxygen, an often-overlooked oxygen consumption load. And the first step produces ionized hydrogen, which can lower the systems pH or consume alkalinity. This gives you an idea of how the biofilter affects other parts of the water treatment process and vice versa.

Both nitrosomonas and nitrobacter are slow growing bacteria. As a result, the acclimation period to develop an active biofilter can take weeks, and sometimes months, before the bacteria populations begin to stabilize. This acclimation period needs to be considered when first starting a farm or at any time when fish loads, and thus ammonia loads, change drastically. This can occur when fish are harvested and then restocked. Similarly, an acclimated biofilter should be kept operating even when fish are not present. This means circulating water and potentially adding an inorganic ammonia source to the water to feed the

Chapter 5: Water Treatment Components

biofilter while waiting for more fish to arrive. In rare cases RAS are operated seasonally, and each year the biofilter must be re-acclimated.

As the biofilm grows on the media it will regularly slough off and into the water. These solids are sometimes consumed by the fish or more often removed by the solids filter. Many biofilter designs encourage the sloughing off of older, dead bacteria by moving the biofilter media around, this allows it to scrape against itself and remove the biofilm via mechanical abrasion. The amount of sloughing off and scraping that occurs determines the biofilm's thickness, which in turn determines the biofilter's efficiency. An optimal film thickness is neither too thick nor to thin. Research is ongoing to determine what biofilm thickness and age is ideal for aquaculture biofilters in both warm and cold water systems.

A range of environmental parameters controls the performance of all biofilters. Biofilms are sensitive to changes in pH with the ideal range of operation being between a pH of 7.0 to 8.0. The pH of an intensive RAS is often below this range due to high concentrations of carbon dioxide from fish respiration (see carbon dioxide stripping section). pH control, also to be discussed in a later section, is primarily for the benefit of the biofilter, as at low pHs the biofilter may cease operating altogether. Similarly, the nitrification rate of a biofilter is also both very temperature and salinity dependent. For example, warm water systems are capable of nitrifying 150% more TAN per square meter as compared to cold water systems, and freshwater systems are capable of nitrifying 100% more TAN

Chapter 5: Water Treatment Components

per square meter than full strength seawater systems. Both factors are an important consideration when calculating the size of the biofilter. The relative ammonia concentration of the water will also affect the nitrification rate, higher concentrations of ammonia means higher rates of nitrification per square meter per day.

It should be noted how nitrogen transformation in an aquaculture biofilter differs from nitrogen removal in a wastewater treatment system. In wastewater treatment, bacteria are used to clean the water in activated sludge reactors. In these reactors heterotrophic bacteria grow rapidly, feeding on the organic carbon sources in the wastewater, incorporating the carbon, nitrogen, and phosphorous that is present in the water into their cells. The bacterial floc is then settled out, and the solids, composed of dead bacteria, are removed. In this process nearly all of the carbon and nitrogen is removed from the water. In aquaculture, autotrophic bacteria transform TAN into nitrate; the nitrogen remains in the water until flushed out during water replacement. In an aquaculture biofilter the growth of heterotrophic bacteria should be limited to increase the space available for autotrophic nitrifying bacteria. Heterotrophs are faster growing and will outcompete autotrophs in certain scenarios, in these cases they will also increase the total oxygen consumption of the biofilter. Limiting organic carbon through solids removal and fine particle filtration (see fine particle filtration section) is the best way to control heterotrophic bacteria.

Chapter 5: Water Treatment Components

Before determining what type of biofilter you would like to use in your system it is important to understand how large it will need to be. Biofilters are sized based on square meters or square feet of active surface area. An approximate surface area can be calculated by making some basic assumptions on the nitrification rate. A typical biofilter nitrification rate ranges from 0.2 – 0.5 g/m^2/day. The exact rate is difficult to calculate and is dependent on a number of factors, most importantly the pH, temperature, TAN concentration, and biofiltration method. Typically higher pH and higher temperatures means a higher rate of nitrification rate. For rough calculations it is best to just focus on temperature and to assume a low nitrification rate for cold water systems and a high rate for warm water systems. Once a nitrification rate has been decided on the next step is to calculate the total surface area needed by dividing the TAN production by the nitrification rate just as we did in the Chapter 3 of this book.

Different types of biofilter media have different specific surface areas, typically written as m^2/m^3 or ft^2/ft^3. The typical range is from 200 – 3000 m^2/m^3. Higher specific surface area means more surface area can fit in a smaller volume, and thus the biofilter has a smaller footprint. A smaller footprint means less building area, less material costs, and easier access for maintenance. However, high specific surface area biofilters come with other considerations, equipment needs, and drawbacks. By dividing the total surface area needed by the specific surface area of the biofilter media you can determine the total volume of media that you require and begin to understand the footprint of the biofilter.

Chapter 5: Water Treatment Components

The final consideration in sizing the biofilter is determining the rate of water flow over the biofilter that is needed to achieve adequate TAN removal. The appropriate steady-state TAN concentration will range from 0 - 2.5 mg/L depending on the species and pH concentration. As mentioned previously in Chapter 4, the pH determines the equilibrium of toxic ammonia to less-toxic ammonium, and different fish species have different tolerances to ammonia concentrations. The other piece of information you need to determine the biofilter flow rate is the biofilter's efficiency, or in other words how much TAN is transformed into nitrite and nitrate each time the water passes through it. Biofilter efficiency rates range from 10-90% and they are dependent on a whole host of factors, most importantly the biofiltration method and temperature. Once you know the TAN produced per day (TAN_{prod}) from chapter 3, and the desired TAN concentration (C_{TAN}) and predicted biofilter efficiency (E_{BIO}) you can calculate the biofilter flow rate as shown in the calculation below.

$$Biofiltration\ flow\ rate = \frac{TAN_{prod}}{C_{TAN} \times E_{BIO}} = \frac{132\ \frac{kg_{TAN}}{day} \times \frac{day}{24\ hr}}{2.5\ \frac{g_{TAN}}{m^3} \times 75\%}$$

$$= 2,933\ \frac{m^3}{hr} = 814\ \frac{liters}{second}$$

If the biofiltration flow rate is lower than the system flow rate that was initially calculated based on the systems ideal hydraulic retention time during the production plan stage, then additional pumping will be needed to recirculate water through the biofilter. This is accomplished by pumping water that exits the biofilter to the entrance of the biofilter, creating a small

Chapter 5: Water Treatment Components

loop of recirculated water. Ideally this does not need to occur, but in systems with larger hydraulic retention times or an especially low requirement for TAN concentrations it may be required.

To be clear, the steps above are a simplified method for properly sizing the biofilter. To arrive at a more accurate number careful consideration needs to go into the systems operating temperature, pH, and TAN concentration, along with the method of biofiltration used. Researchers have discovered a number of equations and constants that properly describe the nitrification rate and efficiency rate of various biofilters. However, even these equations may not properly describe all situations and there is still a degree of guesswork in getting the exact biofilter size correct.

Biofiltration can be accomplished with flowing water over almost any high surface area substrate. However, there are a few methods that are more popular and thus better understood. These include trickle filters, static bed filters, microbead filters, sand filters, and mixed bed bioreactors (MBBR). Lately MBBR has become the standard in large RAS, but all methods have their applicability.

Trickle biofilters have two main components, the substrate and a nozzle. The nozzle distributes water evenly across the top surface of the substrate where water then trickles downwards through the substrate (Figure 6). Common substrates include bioballs (high surface area plastic balls), similarly designed plastic rings, and plastic sheet blocks (sometimes referred to as

Chapter 5: Water Treatment Components

structured media). Structured media is most frequently used in cooling towers on the top of commercial buildings, but it is also a very cost effective substrate for trickle filters. The structured media is sold in rectangular blocks that are composed of rigged sheets that fit together creating a number of channels for water to flow down through. The media comes in a number of different sizes and specific surface areas, the specific surface area can be anywhere between 100-400 m2/m3. Higher specific surface area blocks take up less space, but are more likely to clog with bacteria. Higher water flows will decrease the chance of the media clogging. Trickle filters are typically between 1 m and 2.5 m high, and between 1 m and 3 m wide. The substrate does not need to be cleaned, as dead bacteria will slough off regularly.

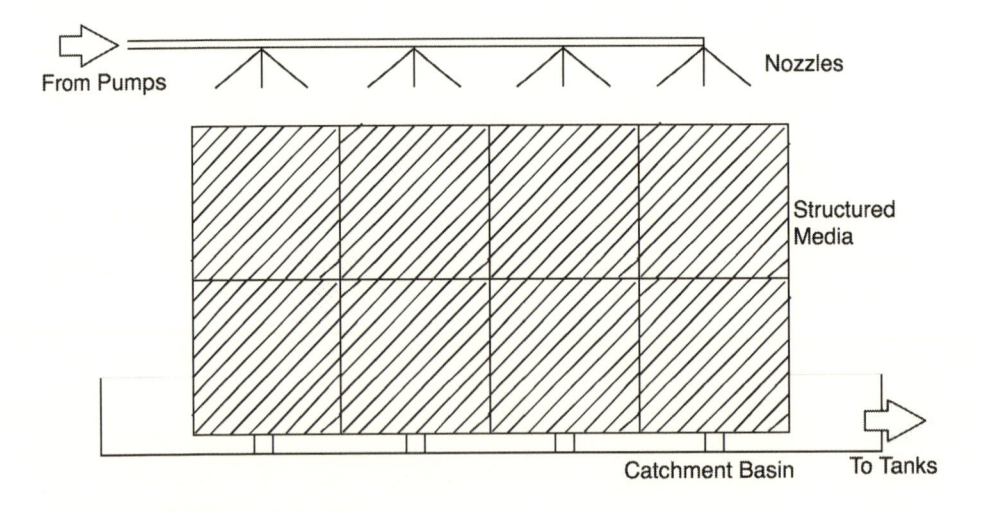

Figure 6. A basic trickle filter utilizing structured media.

The nozzle can be as simple as a PVC pipe with pinholes drilled in it that spray water above the tank, or more engineered with

Chapter 5: Water Treatment Components

specially designed plastic nozzles that create predictable spray patterns. The nozzle's spray pattern is a function of the water flow rate, water pressure, and height of the nozzle above the trickle filter. Manufactured nozzles will have charts to show the spray pattern at various flow rates and pressures. Even spray across the entirety of the substrate will maximize the efficiency of the filter. After the water cascades down through the media it will then either need to be captured in a basin, or in some cases it can fall directly into a fish tank below it. If using a catch basin the basin should be as small as possible and obstructions that could block solids from flowing out of the basin should be avoided.

Many trickle filters also function as carbon dioxide stripping towers, especially when they are coupled with a fan. In fact carbon dioxide stripping towers are often just more loosely packed trickle filters. More discussion on carbon dioxide stripping will be addressed in the next section.

The advantage of trickle filters is that they are cheap and easy to build. Packed media and nozzles are cheap or can even be homemade, and there is no regular cleaning or other maintenance needed. The disadvantages are they require a significant amount of head to raise the water above the filters, they create a lot of spray and will raise the humidity inside the building where they are located, and they are relatively large. The media's specific surface area is not as high as other methods and as result the filters can take up an excessive amount of room. For larger operations this makes the trickle filter impractical, it is best used on small systems. There are

Chapter 5: Water Treatment Components

also issues with system hydraulics when using a trickle filter, the water must be pumped up in order to cascade down through the trickle, then water must flow by gravity through a oxygenation system and into the tanks with enough head to maintain proper tank flow. This requires quiet a bit of head which means the trickle filter must be located at least a couple of meters above all tanks.

Static bed filters are perhaps even simpler than trickle filters. A static bed filter is simply a vessel that contains a nonmoving bed of media that water flows through, upwards, downwards, or sideways. Easy to build static bed filters are made of sand, gravel, or plastic rings/beads. The most important consideration is to make sure water does not short circuit the biofilter media, but instead flows evenly with no dead spots. Although the static beds are easy to build, they are not always easy to maintain. Over time solids will become stuck in the crevices and pathways between the media, clogging the bed and channeling the water flow, thus creating dead spots. To prevent this the beds must be regularly cleaned, this is most easily accomplished by injecting air or water into the bed to fluidize it. Of course, if the media you chose is much denser than water this may require significant energy. Hobbyists frequently use static bed filters and periodically backwash them manually with a shovel or hose.

Bead filters are essentially static bed filters that have been placed inside of a pressurized vessel that can than be easily backwashed. In the solids filtration section we discussed how bead filters can be used for solids removal, but they also serve

Chapter 5: Water Treatment Components

the dual purpose of acting as a biofilter. The high surface area of the beads provides plenty of substrate for bacteria to grow on. When bead filters are operated as biofilters instead of solids filters they are backwashed less often and cannot filter as high of a solids load. They are backwashed less often so that the bacteria populations have a chance to grow and form a biofilm around the media before being scraped off during backwashing. The advantage of bead filters is their dual purpose, and the ability to set them up to backwash automatically with a pressure sensor. However, bead filters need to be prohibitively large to provide enough biofiltration surface area for a large commercial facility, and the larger units are expensive. Furthermore, water must be pumped into the bead filter and the added head of pressurizing the vessel requires larger pumps and more energy. Finally the regular backwashing requires the unit to be taken offline, and water is not being filtered during these times. Consistent filtration is an important part of any farm design especially for a critical process such as biofiltration

In commercial systems the static bed biofilters often do not have a high enough specific surface area and the backwashing needs to occur almost constantly. The next three biofilters we will discuss are all modifications to the static bed biofilter that prevent the need for backwashing by keeping the various media in constant motion.

Sand filters are beds of sand that are liquidized with water flowing through a network of pipes that sit at the bottom of the filter. The sand continuously settles towards the bottom of the

Chapter 5: Water Treatment Components

filter, but upward flowing water that is shot out of nozzles prevents it from ever coming to rest. The bed of sand is in constant motion, and when not in motion the sand is one half to one third its expanded volume. Uniform injection is crucial to maintaining filter efficiency and preventing dead spots or channeling in the sand media. Corners are potential dead spots that need to be avoided, preferably with a round biofilter vessel. The biofilter vessel must also be abrasion resistant, since the sand can erode soft plastic of fiberglass materials over time. Sand media can be purchased in various uniform sizes with specific surface areas between 3000-10000 m^2/m^3, and it will also need to be periodically replaced as sand wears down or flows out of the biofilter. Small sand particles may occasionally overflow from the biofilter if they become coated in a thick biofilm, the reduced specific gravity of biofilms can cause the entire particle to become neutrally buoyant and lift to the top of the water column. A long weir can reduce these losses in the biofilter overflow, as can regular cleaning of the biofilter. At the top of a sand filter there will be a layer of biofilm and other solids that should be regularly removed via a siphon.

Sand filter inlet pipes are typically situated so that they probe downwards into the sand bed with a cleanout port on top of each inlet pipe. This is done to prevent "lockup"; lock up of sand beds can occur when fluidization is stopped for a short time and the sand settles to the bottom. Sand may then backflow into pipes and inlet manifolds. The combination of sand and biofilms can cause the settled bed to turn into a hard concrete like substance. In poorly designed systems the sand can fill pipes at the bottom of the tank and the pumps cannot

Chapter 5: Water Treatment Components

provide enough pressure to purge the sand from the pipe and restart the water motion needed to fluidize the bed. This is easy enough to avoid, but should be considered when designing the inlet manifolds. Pipe cleanouts allow for pressurized water to be injected into each pipe to clean out the sand and restart the flow of water through the pipe. There are many other clever ways to avoid lockup with gravel bases at the bottom of the biofilter or even a vessel that creates a whirlpool like motion of water that can easily be restarted. The easiest method though is pipe cleanouts on all inlets and the prevention of backflow or siphoning in inlets and pipes when water flow ceases.

The advantages of sand filters are the high specific area of sand, up to 10000 m^2/m^3, and the lack of aeration. Also sand media is relatively cheap and filters can be designed for almost any sized system. The drawbacks are the need to continuously fluidize the bed with water. The high specific gravity of sand, which is denser than water, requires a couple meters of head to fluidize and suspend in the water column in a uniform manner. Other drawbacks of sand filters are the potential of the bed to "lock up" and become anoxic if there is an interruption in system power or pumping, and the need for the pumps to be placed directly after the solids filter. Since the sand filter requires the injection of pressurized water to keep it fluidized, the water must then gravity flow through the stripping, oxygenation, and disinfection filtration steps. This limits the options for which methods can be utilized for each of these subsequent filtration steps.

Chapter 5: Water Treatment Components

The microbead biofilter is an inexpensive biofilter option that has been operated at a commercial scale in both fresh and salt warm water systems. The buoyant microbead media is kept below a distribution plate which water flows across before falling down through nozzles that evenly distribute water so that it cascades further down through the media and out of the bottom of the filter. The design of the filter requires the microbeads to be placed inside of discrete cells; the larger the biofilter the greater the number of cells. These cells help with uniformly mixing the media, as water flows down through the buoyant microbeads they move around and scrape against each other.

The microbead biofilter rivals the sand filter in terms of compactness with the media having a specific surface area of approximately 3000 m^2/m^3 for 2 mm diameter beads. Around 40% of the surface area of microbeads is void volume, which is made of small swiss cheese-like channels and chambers that extend towards the center of the bead. The beads are made of inexpensive expanded polystyrene (EPS), the same material used to make Styrofoam packing and coolers. The drawback of microbead filters is that the design, as originally developed at Cornell University, is patented and there has not been a large adoption of the technology within the industry. And although the media is inexpensive, the microbead's vessel, including the distribution plate and cells is complex and costly to construct.

The final type of biofilter to discuss is the mixed bed bioreactor (MBBR), the most widely adopted biofilter for large-scale RAS operations. The MBBR has been adopted by most RAS

Chapter 5: Water Treatment Components

designers and operators because of its reliability, predictability, and low-cost. In an MBBR the water flows into the biofilter vessel that is filled with neutrally buoyant plastic media (Figure 7). The media is agitated by both the water flowing through it and by the airstones distributed throughout the bottom of the vessel. The air is key to preventing dead spots in the biofilter vessel and keeping the media in constant motion. Water flows out from the biofilter vessel over a screened weir that prevents the escape of media.

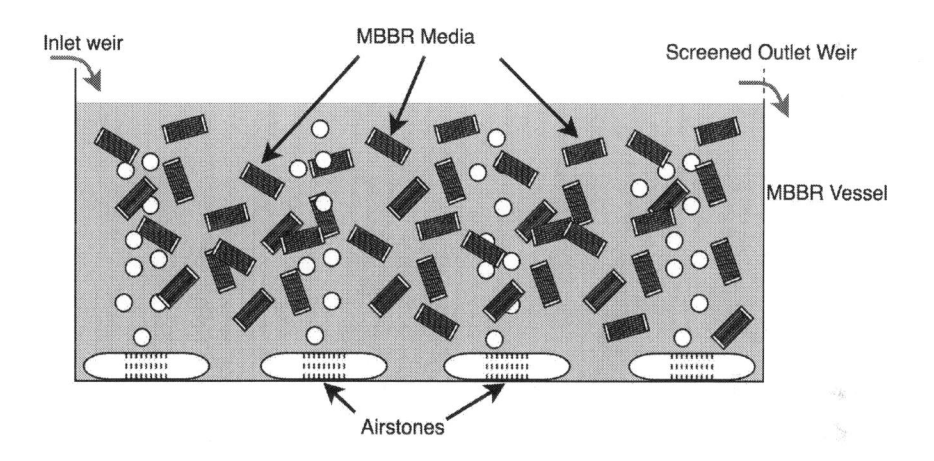

Figure 7. MBBR cross section (not to scale).

There are a number of different vessel options for a MBBR. Small filters can be constructed with closed or open top rectangular and circular plastic tanks, and larger biofilters may be constructed of fiberglass or concrete. The largest biofilters are all typically constructed of concrete, both above and below grade. A typical MBBR is between one to three meters deep and larger units have a rectangular construction (3:1 length to width ratio) whereby water flows from one end of the unit to

Chapter 5: Water Treatment Components

the other, sometimes through separate chambers. The dimensions of the MBBR are flexible and the width, length, and depth can be manipulated to get the desired water volume. The water volume is typically double the volume of the media needed. The vessel then needs to have an outlet or weir that allows water to flow through or over it while preventing the loss of media. Ideally this is done with a screen that has a mesh size slightly smaller than the media being used, that way solids and other particles can still easily pass through, preventing the need for frequent cleaning.

MBBR media comes in a variety of designs and manufacturers tout the benefits of the particular geometry of the media they sell. The truth is it only needs to be neutrally buoyant, non-toxic, low-cost, and have a high specific surface area. Media specific surface areas range from 500 to 3000 m^2/m^3. Higher specific area media will take up less floor space for a given TAN load. The media ranges in price, which is approximately correlated to the specific surface area. When purchasing the media consider the cost of shipping since it may take up a considerable amount of volume in a truck or container. Also choose a media design that has been proven to work in other aquaculture systems.

Aeration can be accomplished with various air disparaging devices such as hoses, stones, or membranes. The aerators should generate medium to large size bubbles that do not require a lot of head pressure, but do create a lot of turbulence in the water column. The goal should be airflow of approximately 7 liters/second per cubic meter of media. The

Chapter 5: Water Treatment Components

benefit of using air in the MBBR is it helps strip carbon dioxide and restores some oxygen to the water. Ideally the head pressure of aeration system is low enough so that blowers can be utilized instead of compressors. Blowers can supply large amounts of air at low to medium heads that will work for most aquaculture applications. There is additional discussion of blowers in the next chapter.

The disadvantages of the MBBR are the need for aeration and the cost of plastic media. Aeration requires blowers that can be expensive and use large amounts of electricity. The plastic media used in MBBRs is also expensive, but it should not require replacement for at least a decade. The MBBR also has many advantages, first it has a low head design and water can gravity flow through it with no need for additional pumps. The MBBR can also be deepened to reduce the biofilters footprint and it can be flexible in its layout. Unlike a sand filter, if the power supply is interrupted the biofilter will be fine and it can easily be restarted. Finally, the system is stable and does not require regular maintenance or special care. The design parameters and operating procedures of the MBBRs are well described for a variety of environmental conditions, reducing the uncertainty of system performance that often comes with biofilters. All of these advantages combine to make MBBRs the right choice for many farmers.

There is a final type of ammonia removal method that cannot be exactly classified as a biofilter. Ion exchange media, such as zeolite, is capable of removing high concentrations of ammonia from water via chemical processes. Zeolite is charged by

Chapter 5: Water Treatment Components

soaking it in a sodium solution, the sodium molecules then occupy the active sites on the surface of zeolite particles. However, once zeolite is placed in an aquaculture environment, these same active sites begin attracting ammonium molecules, which displaces the sodium. This effectively removes the ammonium from the water while adding sodium to it. The media lasts as long as there are still active sites available, and when it fully saturated with ammonium the media can be refreshed with another sodium bath. The downside of this is the sodium bath must occur regularly and the bath water becomes a salt and ammonia filled waste product that is difficult to dispose of. New research has looked into how bacteria can also grow on the zeolite, turning it into a type of biofilter. The high surface area of the ion exchange media looks a lot like a sand filter to bacteria. There is evidence to suggest that the biofilms that grow on the media can also use the ammonium that is captured by the ion exchange media.

The advantage of using an ion exchange media is it can easily handle peaks of ammonia production, whereas a biofilter cannot since it is a living ecosystem that grows slowly to meet demand. At the moment, ion exchange media is not typically used in aquaculture because of the aforementioned cost and hassle of recharging the media. In the future the media may come in handy in specific scenarios, such as long haul live transport, cold water systems, or purge systems, where bacteria growth is either discouraged or too slow to be helpful.

Chapter 5: Water Treatment Components

Carbon Dioxide Stripping

The removal of carbon dioxide from the water is technologically a fairly straightforward process. How carbon dioxide affects both fish health and water quality is not so simple, and requires understanding some basic chemistry. Carbon dioxide is produced by fish at a rate directly proportional to the fish's oxygen consumption, 1.4 kg CO_2 produced per 1 kg O_2 consumed. This comes out to a carbon dioxide production rate of approximately 0.84 kg CO_2 produced per kilogram of feed. As a result the rate of carbon dioxide production goes up when feeding occurs and carbon dioxide levels peak right after eating and then begin to fall until the next feeding event. Once carbon dioxide is dissolved in water it also begins reacting with water to form carbonic acid.

$$CO_2 + H_2O \leftrightarrow H_2CO_3^*$$

It is difficult to determine how much carbon dioxide is dissolved in the water and how much is in the form of carbonic acid since this reaction can rapidly go back and forth depending on the equilibrium conditions. As a result you will often see carbonic acid and carbon dioxide measured together as $H_2CO_3^*$, with the asterisk indicating it is present in both forms, although in most cases the majority is in the dissolved carbon dioxide form. Carbon dioxide when in the form of carbonic acid is a part of the total carbonate carbon system. Total carbonate carbon is composed of three forms, carbonic acid ($H_2CO_3^*$), carbonate (HCO_3^-), and bicarbonate (CO_3^{-2}). The pH of the system dictates the ratio of each form (Figure 8).

Chapter 5: Water Treatment Components

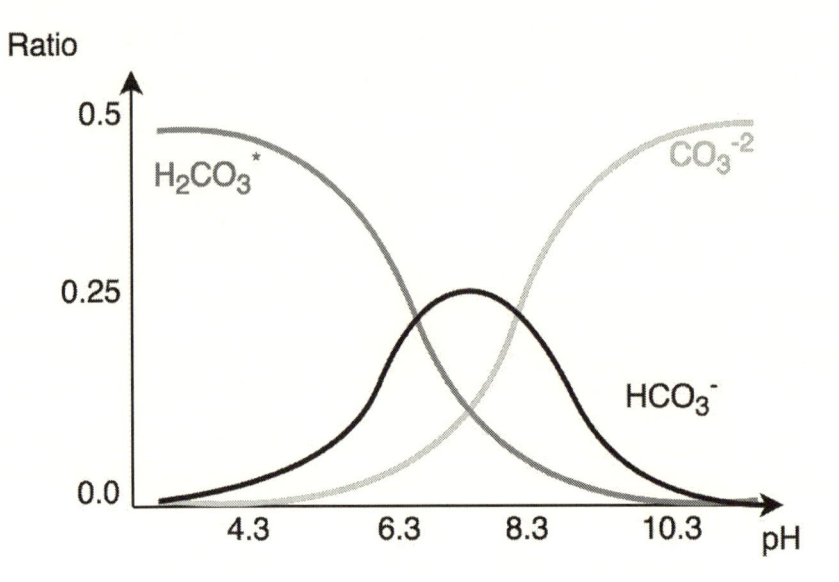

Figure 8. Ratios of carbonate carbon species as a function of system pH.

The above chart shows you how carbon dioxide is tied directly to system pH. As the pH changes so do the ratios of the total carbonate carbon system, and similarly if we change the ratios of the carbonate system, say by adding carbonic acid to the water, we will also affect the systems pH. The carbonate system and pH are also closely related to the alkalinity of the system water, this is discussed in more detail in a following section. For the moment we just need to know that carbon dioxide in high enough concentrations can lower the pH and this can be harmful to fish.

Tolerance levels for exactly how much carbon dioxide fish can handle for extended periods of time is still up for debate. New guidelines suggest that dissolved carbon dioxide levels of 10-25 mg/L are appropriate for most fish species. In natural waters

Chapter 5: Water Treatment Components

where carbon dioxide is in equilibrium with the environment the concentration is around 1 mg/L. High carbon dioxide levels and low pH lead to what is known as both the Bohr effect and Root effect. The Bohr/Root effect lowers the capacity of the fish's blood to carry oxygen. The reduced oxygen carrying capabilities puts significant stress on the fish. Given all of this information it makes sense to expend a considerable amount of energy to remove carbon dioxide from the system since it will lead to improved fish health and faster growth. Carbon dioxide removal was neglected in many early RAS, but as stocking densities and feeding rates have increased it has become an important water treatment step.

The stripping rates of various treatment methods is difficult to measure and currently there is no standard that can be used to compare various methods against each other. The starting and ending concentrations of carbon dioxide, water temperature, elevation, alkalinity, and pH will all influence the rate of stripping. The carbon dioxide concentration of the air inside of the farm building is also a factor, often times elevated levels of carbon dioxide are found inside farms due to the large amount of respiration that is occurring in an enclosed space.

The standard aeration efficiency (SAE), which measures the efficiency of various devices to oxygenate the water, can also be used to approximate the stripping efficiency. Under normal aquaculture conditions the stripping efficiency of a given aeration device will be 80-90% of the SAE. For many methods of aeration this gives a rate of around 2.5 kgCO_2/Kw-H. Carbon dioxide is difficult to remove from water because of a low

Chapter 5: Water Treatment Components

molecular diffusivity rate at normal aquaculture conditions, its interaction with the carbonate system, and its high solubility ratio in water. It behaves very differently from oxygen, the other major gas that we have to deal with in aquaculture. As a result, significant airflow and thus energy is needed to remove carbon dioxide from the water. The good news is that carbon dioxide is stripped whenever water comes into contact with air. This means an MBBR helps with carbon dioxide stripping, as do any weirs located in the system, and even the water surface area of the tanks helps strip carbon dioxide. However, all of this stripping should only be considered a bonus, and can be ignored for the purposes of calculating how much total stripping is needed.

Given the lack of available standards, there is debate over the most efficient method to strip carbon dioxide and many designers use different methods. Common methods include aeration towers, surface aerators, and submerged diffusers of various designs. Despite a lack of agreement on the stripping method there are some basic rules to follow when designing for each of the possibilities.

Aeration towers, also known as packed columns, consist of water dripping through a distribution plate or being sprayed out of a nozzle before then cascading down through a tower, which is sometimes filled with media to further break up the water. A key piece of aeration towers are fans that move air through the tower. Generally an airflow to water flow ratio of 10:1 is required to maximize stripping efficiency. The large volume of air increases the concentration gradient of carbon

Chapter 5: Water Treatment Components

dioxide between air and water, leading to quicker diffusivity of carbon dioxide from water to air. Similarly, it is best to operate the inside of the tower under a slight vacuum by using the fans to push air out of the tower. This slightly lowers the saturation concentration of carbon dioxide in the water when in equilibrium with the lower pressure air. The fans used in aeration towers need to be able to be able to move air at a small pressure, typically measured in inches of water. Assume a pressure of 0.5"- 1" for most aeration towers. Fans that would typically be used in high moisture HVAC environments are appropriate for use in aquaculture. The tower is best designed so that water and air flow counter current to each other. This maximizes the concentration gradient between the fluids.

As mentioned many towers have media inside of them to break up the water flow and create a certain level of air turbulence and mixing. This can look similar to a trickle filter, and there are clever designs that combine the carbon dioxide stripping function of an aeration tower with a trickle filter. The challenge comes in matching the sizing for both water treatment steps. Additional biofiltration or stripping may be needed if the tower is undersized for one or the other. Ring shaped media with significant void space are often used in aeration towers as is packed column media. Any media may foul over time, especially if the water flow rate is too low to knock off the existing biofilm. Prefabricated aeration towers are sold by some equipment manufacturers, but they can also be self-fabricated out of welded aluminum or repurposed plastic barrels. When using the tower it is important to ensure that there is enough head pressure after the tower to be able to flow through the

Chapter 5: Water Treatment Components

oxygenation system and back into the tanks. The downside of aeration towers is the fact that water must be pumped up a few meters in order to be able to cascade down through the tower and then gravity flow back to the tanks. Other stripping methods do not require as much head pressure and allow for pumping to take place after the carbon dioxide stripping treatment. The benefit of aeration towers is that they do a good job of carbon dioxide stripping and have a high rate of stripping efficiency if designed properly.

Surface aerators are a low-head option for carbon dioxide stripping. There are a number of different surface aerator designs; common ones either employ a propeller spinning at high speeds just below the surface of the water or a paddlewheel spinning at high speeds striking the surface of the water. Both designs aim to stir up the surface water into a white froth that typically sprays about 1 meter high and a couple meters wide. Surface aerators are very common in pond aquaculture to oxygenate the water. They can be considered low-head even though no actual pumping occurs because the head is equivalent to approximately how high they toss the water up in the air, usually only a meter or so.

There are many surface aerator manufactures, and each will claim that theirs is the most efficient; the truth is they all have approximately the same efficiency. The largest units are around 2.2 Kw and multiple units may be needed for a large system. The drawbacks of surface aerators are that they are another motor that requires maintenance, and shields may be needed to prevent overspray from leaving the system or getting other

Chapter 5: Water Treatment Components

equipment wet. Their use in RAS is relatively new compared to aeration towers and more time is needed to see if they are viable carbon dioxide stripping method.

The final option for carbon dioxide stripping is submerged aeration, sometimes just called aeration. It involves the use of air stones, hoses, or other membranes to create bubbles at the bottom of the tank, same as what is needed in an MBBR. A blower is needed to provide enough pressure and flow to adequately aerate the water. Generally, an airflow to water flow ratio between 3:1 to 6:1 is ideal. Aeration basins are typically between 2 m and 3 m deep. Equipment manufactures offer a variety of choices that will create an even bubble pattern from pressurized error. Again, all of them will tell you that theirs is the best and most efficient method to create small bubbles for the least amount of energy expenditure. It is difficult to make comparisons to prove which bubble maker is best, however physics says that creating small bubbles requires some amount of energy and there is no way to get around this. As a result their efficiencies for carbon dioxide stripping purposes are all approximately equivalent. There is some debate as to what bubble size is best for carbon dioxide stripping. Larger bubbles require less head pressure, but they create less surface area for diffusion of gases. Small bubbles require more head pressure, but create more surface area. No consensus has been reached on what works best, but medium bubbles, approximately 1mm in diameter, are typically used in practice.

The advantage of submerged aeration is that the basin can be located immediately after the biofilter, requiring no additional

Chapter 5: Water Treatment Components

pumping head. The drawback is the stones or membranes that created the bubbles may need to be cleaned periodically, and a blower is needed for airflow. A challenge with all carbon dioxide stripping methods is that most are only 40-65% efficient per pass of water. This means that a lot of water needs to be pumped in order to properly strip the water and in fact this can become the determining factor in choosing a water flow rate in high-density systems. A multistage aeration basin is one solution to increase efficiency and future work may look at how vacuum equipment can pull carbon dioxide from the water. Carbon dioxide stripping practices are still being refined, but intensive RAS requires a properly sized treatment step for removing carbon dioxide from the water. Aeration basins currently offer the best solution for a low head stripping system.

Disinfection

Disinfection is needed to lower the population of potential pathogens in the system. One hundred percent eradication of bacteria, protozoa, and viruses is never possible, but by keeping their populations as low as possible the threat of disease outbreak is significantly reduced. The amount of disinfection required will depend on what kinds of potential pathogens need to be combated. Viruses are generally more resistant than bacteria and subsequently require higher doses of a given disinfectant to be eradicated. In most industries disinfection is accomplished chemically with chlorine, iodine, peroxide or alcohol based solutions. However these chemicals are not considered fish safe and are difficult to administer in

Chapter 5: Water Treatment Components

large enough quantities to disinfect an entire RAS. As a result, there are two widely adopted methods that are appropriate for aquaculture disinfection, Ultraviolet light (UV) and Ozone (O_3). Each method has its advantages and drawbacks, and sometimes they are used in tandem. Also, ozone has uses outside of pure disinfection that will be discussed in the fine particle filtration section.

UV treatment equipment consists of water flowing over tubular light bulbs encased in quartz sleeves. The UV bulbs are designed to emit light at a wavelength of approximately 250 nanometers (nm). The UV radiation band is between 190-400 nm, with 262 nm being the most destructive to biological life. UV dosages are measured in either millijoules per centimeter square (mJ/cm^2) or joules per meter square (J/m^2), note that 1 mJ/cm^2 is equal to 10 J/m^2. A minimum dosage of 30 mJ/cm^2 is recommended with dosages regularly increasing to 180 mJ/cm^2 with an absolute maximum dosage of 300 mJ/cm^2.

In order to properly size a UV filter, the water flow rate, water transmittance, and desired dosage must be known. The flow rate will typically be dictated by the flow rate of the system since it is best to treat all the water and not just a side stream. The transmittance of the water is determined by the clarity of the water and the amount of solids in it, transmittance of 70-90% is normal in most aquaculture systems. Lower transmittance means dirtier water, and more power is required to supply the desired dosage. As a result, removing all solids before disinfection by UV is key to maximizing the effectiveness of the system. The dosage will be determined by what

Chapter 5: Water Treatment Components

pathogens are of the most concern for the species you are growing. For example, vannamei shrimp are sensitive to various viruses that require a higher dosage to kill. These three parameters determine the amount of power, as measured in watts (W) that is needed for the system. UV lamps are commonly available in wattages from 50 to 320 W. Multiple lamps are often used in a single unit to provide enough power to properly disinfect the water. As lamps age they slowly become less effective and the power they deliver at the optimal wavelength is diminished. The best UV lamps have a lifespan of 12,000 hours and as a result they need to be replaced once every 500 days. Similarly the quartz sleeves that encase the bulbs need to be cleaned or replaced regularly in order to maximize their transmittance.

There are two major types of UV filters: pressurized and open channel. A pressurized UV filter is typically a pipe or another vessel that is plumbed in line with the system piping and pressurized by the system pumps. Pressurized filters are most often made of either stainless steel or UV resistant plastics. The head loss across the pressurized vessel can be significant for some designs, especially when operated at the limit of their designed flow rate. Generally, lower dosages means higher flow rates and higher head losses for a given unit. Open channel UV filters, on the other hand, are low-head as water simply flows through a channel that is filled with vertically mounted UV bulbs. The open channel design works well for large flow rates above 12,000 LPM. Constructing a channel though is costly and takes up significant room. Open channel UV filters are rare in current RAS facilities, but they have proven

Chapter 5: Water Treatment Components

to work well in large fish hatcheries and in wastewater treatment facilities. As mentioned, all UV filters require regular cleaning to ensure a high transmittance rate through the quartz sleeve. One advantage is the UV light makes it very difficult for bacteria or algae to grow on the inside of the filter, but things still get dirty over time. Large UV filters typically include a cleaning apparatus, sometimes called a wiper, that allows the sleeves to be brushed clean while keeping the unit online, if this is not provided as standard it should be purchased to make the cleaning easy and accessible.

The advantages of UV filters are their plug and play ease of use and lack of need for continuous monitoring and alarm sensors. They can be designed and operated at low head and they will kill all pathogens if a high enough dosage rate is used. The disadvantages are the need for clear, high transmittance water, regular cleaning of sleeves, and regular replacement of lamps. Fortunately, the maintenance can be easily scheduled, and the water should be kept at a high enough transmittance regardless to ensure fish health. Extremely high flow rates that large scale RAS requires can be difficult to treat at high enough dosages with off-the-shelf equipment. Custom designed and expensive open channel UV filters are needed for the largest farms. Hopefully in the future off-the-shelf components for open channel UV will lower the costs.

The second method of disinfection treatment in aquaculture systems is ozone contactors. Ozone is an oxidizing agent that reacts with all organic compounds, and as a result it is very effective at killing pathogens. The reason ozone can be used in

Chapter 5: Water Treatment Components

aquaculture systems is it decays rapidly back into oxygen, essentially becoming inert before it is returned back to the fish thanks. However, using ozone for disinfection is not foolproof and requires careful monitoring and a number of material and safety considerations.

The ozone dosage is determined by the concentration of ozone and the contact time. Ozone has uses beyond simple disinfection, as it can also oxidize nitrite into nitrate and it decomposes some fine solids. As a result, the amount of ozone needed for disinfection is dependent on the amount of all organic and inorganic compounds, like nitrite, in the water. A large amount of organic compounds will use up all of the ozone and allow some pathogens to pass through unscathed. The ideal dosage is between 0.1-1.0 mg/L with one to ten minutes of contact time. A lower concentration of ozone is needed for longer contact times and a higher concentration is needed for shorter contact times. In drinking water treatment a dosage of 650 mv ($\approx$0.25 mgO$_3$/L) for thirty seconds is recommended to kill most harmful bacteria. However, obtaining these dosages and contact times in large scale RAS is difficult given the amount of water that is recirculated and the large amount of organic compounds in the water.

Ozone levels are monitored by measuring the oxidation-reduction potential (ORP) of the water with an electronic sensor. The ORP measures the tendency of electrons to be accepted by an oxidizing agent in solution (ozone in this case), thereby being reduced (gaining electrons). ORP sensors typically have units in millivolts (mV) or volts (V), which can be

Chapter 5: Water Treatment Components

converted into units of milligrams of ozone per liter (mg/L) if temperature and pH are also known. Also it should be noted that most ORP sensors do not detect below a limit of 0.1 mg/L of ozone. In order to safely use ozone in an aquaculture system the residual dosage that is present after the contactor must be continuously monitored with an ORP sensor and automated feedback must be used to increase or decrease the dosage as needed. Ideally a residual dosage that is barely detectable is present after the contactor with the hopes that it is used up completely by the time it returns to the tanks.

Ozone can be injected into the system inline through a large pipe or through the use of a dedicated ozone contactor. A contactor should be designed to have enough volume to ensure that the water has adequate contact time with the ozone solution before exiting. Off the shelf, large-scale ozone contactors and generators are expensive and currently not optimized for aquaculture. Smaller side stream contactors are available, but these cannot be used for complete water disinfection. Ozone is typically delivered into the water through the use of a venturi nozzle. A venturi operates by pumping water through a cleverly designed nozzle that creates a lower pressure zone whereby gas is sucked into the water stream and mixed in as small bubbles. The advantage of using a venturi for ozone is that it creates a vacuum operated delivery system rather than a pressurized delivery system, reducing the chances of leaks into the atmosphere. The venturi nozzle is an efficient method for dissolving gases into water but requires the use of a dedicated pump to push water through the nozzle. Ozone is made with an ozone generator. The generator is fed pure

Chapter 5: Water Treatment Components

oxygen gas that is converted into an ozone and oxygen gas mixture that can then be delivered into the water. More discussion is given to ozone generators in a following section. All ozone systems and parts need to use corrosion-resistant materials as ozone can oxidize plastics, rubbers, or non-corrosion resistant metals.

The advantage of ozone is it has multiple uses beyond disinfection. However, the disadvantages are that it is hard to scale in order to provide adequate disinfection in large systems, and it requires continuous monitoring and control. It is difficult to consistently deliver the correct dosage into the water so that it can provide complete disinfection. Once water leaves the contactor, it needs to have an ozone level near zero or below the detectable limit. Maintaining a high enough dosage inside of the contactor and insuring it decomposes soon after exiting can be a difficult task. In order to ensure fish are never exposed to ozone the water can pass through a de-ozonation contact chamber, although this is rarely viable. Another option is to locate the ozone injection point before the biofilter, this way any residual ozone will be consumed by the organic material in the biofilter. A final option is to dose the water with UV radiation since it has the ability to hasten the decomposition of ozone back into oxygen. However, dosages of at least 50 mJ/cm^2 are needed to adequately decompose the ozone and that point you might as well just be using UV to do all of the disinfection. An additional drawback is the ability of ozone to potentially create toxic bromide compounds in saltwater systems; given this possibility it is best to avoid mixing saltwater and ozone. Finally, ozone at concentrations in

Chapter 5: Water Treatment Components

the air above 100 parts per billion is harmful to humans. Whenever ozone is present alarming sensors should be used to monitor the air quality around ozone generators and ozone injection locations. Similarly, ozone equipment needs to be regularly checked for leaks or signs of corrosion. Given these challenges, ozone is not recommended for use as a disinfectant on large-scale systems, however it does play an important role in most fine particle filtration systems, which will be discussed in a following section.

Disinfection is needed in all aquaculture systems to beat back the threat of pathogens and the potential of disease outbreak. In small-scale systems there are a number of equipment options for both pressurized UV filters and ozone contactors. In large-scale systems with high flow rates above 12,000 LPM only the largest pressurized UV filters and open-channel UV filters are an option. Other UV options or ozone contactors can only be operated as side streams in the larger systems.

Oxygenation

Oxygenation is necessary in all RAS to provide adequate oxygen for fish respiration at higher stocking densities. The total oxygen needs of the farm are directly related to the amount of feed consumed by the fish as shown in the calculation performed in the production plan section. Approximately 0.6 kilograms of oxygen are consumed per kilogram of feed. Once the total oxygen needs of the farm are calculated the method of oxygenation can be chosen. Different methods have different levels of efficiencies and their

Chapter 5: Water Treatment Components

effectiveness is highly dependent on other water quality parameters. Remember that the saturation concentration of oxygen is between 8-10 mg/L and is mostly dependent on water temperature and salinity, with higher temperatures and salinities lowering the saturation concentration. Oxygen levels need to be maintained above 5 mg/L to eliminate fish stress due to low oxygen. Oxygenation methods can be operated at a variety of pressures, which changes the saturation concentration and greatly affects their efficiency. The actual oxygen that is used in the oxygenators can be supplied by either a liquid oxygen tank or an oxygen generator; each of these sources is discussed in the next chapter.

The terms aeration and oxygenation are often confused or used interchangeably. However, aeration refers to the use of air to raise the water's oxygen concentration, while oxygenation refers to the use of pure oxygen to accomplish the same. Aeration is not viable for intensive aquaculture since the maximum stocking density that can be achieved for a low-oxygen tolerant species with the use of a lot of air is 40 kg/m^3. This is for the simple reason that the saturation concentration of water at atmospheric pressure is between 8.0 and 10 mgO$_2$/L for most aquaculture appropriate water temperatures. Aeration cannot raise the oxygen levels above this concentration and maintaining concentrations at 6 mgO$_2$/L requires a lot of energy.

Aeration energy efficiencies are approximately 0.2 kgO$_2$/kW-h under normal aquaculture conditions, where as oxygenation energy efficiencies are typically around 1.0 kgO$_2$/kW-h.

Chapter 5: Water Treatment Components

Typically you will see a standard aeration efficiency (SAE) of between 1.5 - 2.5 kgO$_2$/kW-h for most commercially available aerators. SAE is great for comparing one aeration method to another, however it is not accurate for determining what efficiencies will be achieved in a normal aquaculture system. This is because SAE is calculated by using a specific set of standard conditions where the beginning concentration of oxygen in the water is 0 mgO$_2$/L, which is far below the oxygen concentrations found in any aquaculture system.

$$\frac{dC}{dt} = K_{LA}(C^* - C_0)$$

This equation shows how the rate of change in oxygen concentrations (dC/dt) is equal to the aerator transfer efficiency (K$_{LA}$) multiplied by the saturation concentration (C*) minus the initial concentration (C$_0$). Thus the rate of change is maximized when the initial concentration is zero. K$_{LA}$ is specific for each aeration method and it is related to the surface contact area of the air and water generated by the aerator along with other factors. It can be determined experimentally by tracking the change in oxygen levels over time in a batch reactor.

Aeration is appropriate for lower stocking densities (<25kg/m^3) or in purging or holding systems where there is no feed entering the system. In these cases aeration can be accomplished with a variety of methods, all of which are very similar to with what is used to perform carbon dioxide stripping. The most common aeration method is the use of airstones and a blower. The bubble size will determine the

Chapter 5: Water Treatment Components

oxygen transfer efficiency of the aerator. The airstones are typically placed on the bottom of the tank and a blower supplies the air. Airstones come in a variety of shapes and materials, but all are equally good at getting air into the water. Other aeration methods include surface aerators and aeration towers, same as those used in carbon dioxide stripping. The exact sizing of the aeration system will depend on the oxygen consumption of the fish, and its efficiency will depend on the water temperature, salinity, and the oxygen concentration that you wish to maintain. The oxygen adsorption efficiency of most aerators is 10% or less under ideal conditions. Aeration is a good option for small systems with low oxygen, where the cost of an oxygenation system is not economically feasible. In most of these systems aeration with airstones and a blower is probably the best option.

In contrast to aeration, oxygenation has the ability to significantly raise the oxygen levels in the water above the saturation concentration at normal atmospheric conditions. This is because oxygenation uses "pure" oxygen, with an oxygen concentration above 90%. Henry's law states that the amount of dissolved gas is proportional to the gases partial pressure. Higher partial pressures of oxygen creates a new equilibrium according to Henry's law, which means that C^* is much higher when pure oxygen is used. Instead of the saturation concentration being 8-10 mg/L it is now 40-50 mg/L. Furthermore we can inject the oxygen into the water at pressures well above atmospheric pressure. This raises saturation concentration to above 100 mg/L. This means a lot of oxygen can be added to the water in a short period of time,

Chapter 5: Water Treatment Components

allowing for more fish to be supported per liter of water. We typically calculate oxygen consumption in kilograms per hour or per day but a lot of the oxygenation equipment is rated in liters per minute (LPM) of oxygen delivered. The density of oxygen depends on the pressure of the gas. Liquid oxygen has a density of 1.14 kg/L, whereas gaseous oxygen at 0.7 bar has a density of 0.0022 kg/L. Thus 7.6 LPM of oxygen would be needed to deliver 1 kilogram of oxygen per hour to the water. A variety of methods can be used to deliver this oxygen into the water, all work more or less with the same operating principle of maximizing the contact surface area and contact time of the gaseous oxygen with the water.

There are four common methods of oxygenation in RAS: oxygen cones, LHOs, u-tubes, and diffuser stones. The oxygen cone is typically a fiberglass cone-shaped vessel where water enters the top of the cone and flows out the bottom (Figure 9). Oxygen is injected into the cone through an orifice, whereby bubbles rise against the downward flow of water. The water velocity inside of the cone decreases as you move down the cone. This shape allows for all bubble sizes to remain in suspension inside of the cone until they are completely dissolved. The rising bubble cannot leave the cone at the top due to the high water velocity and it dissolves completely before it can be flushed out the bottom of the cone. This design ensures that under the correct operating conditions all injected oxygen is dissolved into the water.

Chapter 5: Water Treatment Components

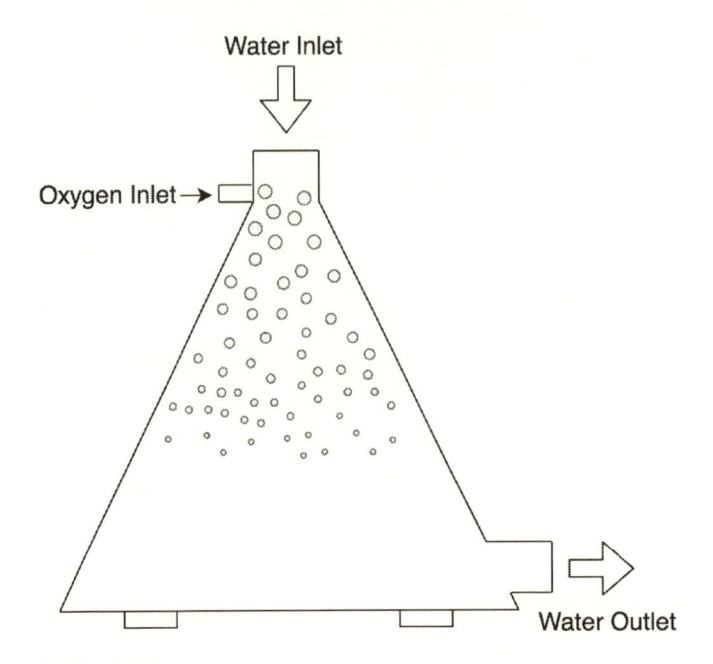

Figure 9. Oxygen cone bubble size distribution.

The largest single cone design is able to handle flow rates of approximately 6000 LPM, while the smallest cones can handle 100 LPM. A cone can be operated either as a side stream or as a part of the main treatment flow. There are trade-offs for each method. As a side stream, a separate side stream pump can supply a cone with a small water flow rate at very high pressures, this allows for a lot of oxygen to be dissolved with a smaller sized cone. The other option is to use multiple large cones on the main flow operated at lower pressures. Both options can supply the same amount of oxygen, but under different flow rates and pressures. The lower energy needs of the main treatment flow make it the better option even though the capital costs and space needed are higher.

Chapter 5: Water Treatment Components

Oxygen cones are always operated under pressure to increase the dissolved oxygen saturation concentration; the pressure typically ranges from 3 to 30 meters of head (approximately 0.3 to 3.0 bar). A quick aside on pressures, when we talk about meters of head we are referring to meters of water, this is the pressure generated by that depth of water. Pressure is also commonly measured in bars, atmospheres (atm), pounds per square inch (psi), and pascals (Pa). For the purposes of this book we will typically refer to pressure in meters of head (of water) since this is a more practical unit that can be easily visualized and understood. Also note that all pressures are relative to atmospheric pressure, which is 10.3m or 1 bar. In the context of oxygen cones, pressures below 7 meters of head would generally be considered low pressure with greater pressures being high pressure. The pressure inside of the cone is the pressure of the water created by the pump. The pressure is raised inside of the cone by choking a valve on the outlet of the cone. A cone should always be operated with a manual or automated pressure gauge and control. Similarly, the oxygen feed should have an adjustable flow gauge.

The pressure, water flow rate, and oxygen flow rate will determine the dissolution efficiency of the cone. The dissolution efficiency is the percent of oxygen that dissolves in the water after being injected into the cone. The efficiency will be low if the oxygen cone pressure is low and a lot of oxygen is being injected into the cone. Not all of the gaseous oxygen will dissolve into solution and some will bubble out into the air. Ideally the cone should be operated such that the efficiency will be maintained above 90%. The trade-off is that more oxygen

Chapter 5: Water Treatment Components

can be dissolved into the water in a smaller cone at lower dissolution efficiencies, however you are also wasting oxygen. Oxygen cones are sized based on the manufacturers specifications. A desired flow rate and pressure are selected, which will help determine the size of the cone and the amount of oxygen that can be dissolved at a efficiency of 90% or greater.

The use of new low-pressure cones and other shaped vessels is beginning to be explored with some successful commercial operations now using low-pressure vessels. These vessels operate at one to three meters of head and require less energy than traditional cones. The vessel can be positioned below grade such that oxygen is injected a couple meters below the water level of the tanks. This means that oxygen can be injected under pressure without having to choke a valve to create artificial pressure. In other words, the two meters of head inside of the vessel is not two meters of head that the pumps must work against thereby saving some pumping energy.

The advantages of a cone are they are easy to operate, relatively inexpensive, and don't require a lot of space. A high-pressure cone can inject up to thirty times the oxygen that would be available under normal atmospheric conditions into the water, significantly increasing the carrying capacity of the RAS. The downside of cones is that maintaining the water pressure inside of the cone requires significant pumping energy. Another disadvantage is that if operated as a side stream only, the side stream water is oxygenated and it must

Chapter 5: Water Treatment Components

be adequately mixed into the main treatment flow to evenly distribute the oxygen throughout the system. Other lower energy options are available for oxygenation.

The low-head oxygenator (LHO) is a relatively recent invention and is now the second most common method of oxygenation in RAS. As its name suggests, the LHO uses less head and thus less pressure to inject oxygen into the water. An LHO consists of a orifice plate that contains multiple 0.5 to 1 cm diameter holes evenly distributed approximately every 10 cm across it. Water flows over the orifice plate and drips through into a chambered vessel that is filled with pure oxygen gas (Figure 10). The water droplets pass through the oxygen gas and land at the bottom of the LHO where they then flow out into the tank or the next water treatment step. An LHO only has as much pressure inside of it as the water depth sitting on top of the orifice plate, usually 0.1 to 0.3 meters. Despite this low pressure the pure oxygen environment inside of the LHO allows for a significant amount of oxygen to diffuse into the water. Typical dissolved oxygen concentrations leaving an LHO are around 15.0 mg/L. The LHOs design can be scaled to accommodate any sized water flow; larger water flows require a larger surface area of orifice plates.

Chapter 5: Water Treatment Components

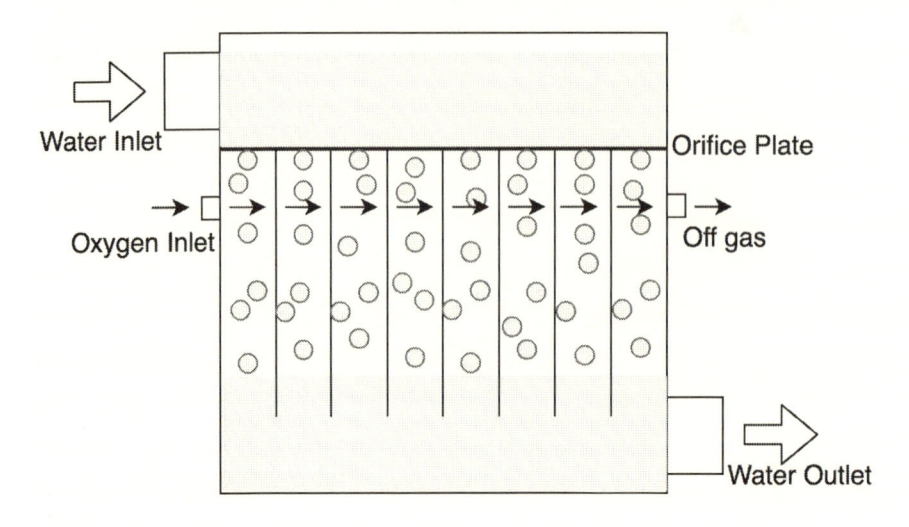

Figure 10. Cross-section of an eight chambered low-head oxygenator (LHO).

The chambered design allows for a higher rate of oxygen diffusion into the water. This is because the gas inside of the LHO as inside of any oxygenator is made up of not only oxygen but also the nitrogen and carbon dioxide that diffuses out of the water into the gas inside of the oxygenator. As stated in Henry's law, both the gas and liquid must be in equilibrium, this means oxygen goes into the water and nitrogen/ carbon dioxide comes out until the relative proportions in each are equal. The chambered design takes advantage of this fact to decrease the proportion of oxygen left in the last chamber before it is off-gassed. Since the LHO operates at lower pressures not all of the gas that enters the LHO is dissolved into the water. There is still a gas mixture of oxygen, nitrogen, and carbon dioxide left that must be off-gassed to prevent the buildup of gases inside of the LHO. In the first chamber the proportion of oxygen is near 100% but as the gas moves

Chapter 5: Water Treatment Components

between chambers through a drilled hole it slowly becomes approximately 25% oxygen. This means that LHO efficiencies are not nearly as high as oxygen cones. Instead the dissolution efficiency of an LHO is usually around 60 to 70%.

LHOs are commonly custom made and sized for each operation; common construction materials include aluminum and plastic. The proper number of chambers, orifice size, orifice distribution, and over all size have all been researched and empirically determined. Ask an experienced engineer or manufacturer to help you determine the best LHO sizing for your operation. The advantage of an LHO is it can be operated with gravity flow and does not require the main treatment pumps to push more head for oxygenation. LHOs can handle very large water flows and can evenly distribute oxygen into the main treatment flow. The drawback is that they have lower efficiencies and a lot of oxygen is wasted and not dissolved into the water. Also the off-gassing can be difficult to operate. The off-gas valve should maintain a constant pressure inside of the LHO, such that as the pressure builds excess gas is purged. Gas purging valves can easily foul and may require regular adjustment. Finally, the maximum oxygen concentration achieved by LHOs is relatively low compared to other higher-pressure methods and thus the carrying capacity of the system is limited to the water turnover rate.

A less common low energy method of oxygenating water is through the use of U-tubes. U-tubes are large pipes that run 3 or 5 meters below grade and then return back to grade. The oxygen is bubbled in with an airstone or other diffuser near the

Chapter 5: Water Treatment Components

bottom of the U (Figure 11). The pressure where the oxygen enters the water is equal to the water depth of 3 to 5 meters. The oxygen bubbles rise up the water column, and under the correct conditions when there is a high water flow rate and low oxygen needs, the efficiency can be nearly 100%. Higher dissolved oxygen concentrations require more oxygen to be bubbled into the water and as a result there are lower efficiencies of 50-60%. The U-tube is really just a method of creating a deep-water column where oxygen can be injected at depth, thereby increasing its saturation concentration. The water can gravity-flow through the U with minimal head.

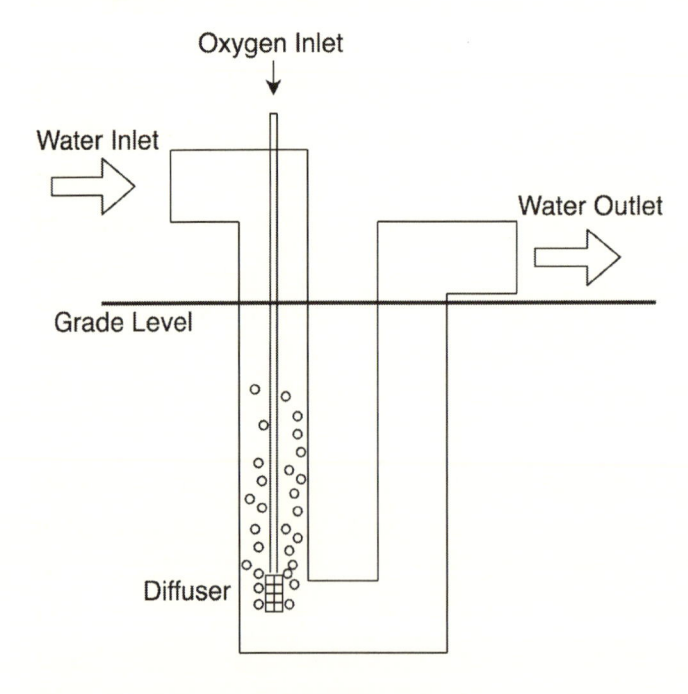

Figure 11. U-tube cross-section.

Chapter 5: Water Treatment Components

The construction of U-tubes is relatively easy if it is possible to dig deep into the soil below the farm. However, bedrock and rocky soils can make the excavation prohibitively expensive or near impossible. Evaluate your local soil conditions to determine if U-tubes are even an option. The diameter of the pipe that is used to make the U-tube will depend on the water volume that needs to be passed through it. Multiple U-tubes can be used in parallel to handle large flows. The height difference between the inlet and outlet of the U-tube should be enough to overcome the head losses of water flowing through the pipe. The oxygen should be injected on the downward flowing section of the U-tube with a fine bubble diffuser. The fine bubble diffuser is best lowered from the top of the tube so that it can be easily retrieved and replaced if needed. The advantage of U-tubes is they are foolproof and reliable. They also require very little head and do not need to be pressurized by a pump. The drawback is they can be expensive or impossible to construct in some areas, and they have middling efficiencies if high dissolved oxygen concentrations are needed.

The final oxygenation method is to use fine bubble diffusers supplied with pure oxygen inside of tanks and sumps. This is by far the least capital-intensive choice and also the least efficient with oxygen dissolution efficiencies of approximately 40% in the best case scenario. Diffusers are good for temporary systems or as a stopgap solution in the case of emergency (see emergency oxygen section). However, as the main, permanent oxygen solution it is not viable since the efficiency is so low. If diffusers are used they should be located in the deepest tank

Chapter 5: Water Treatment Components

or sump possible and should be placed as near as possible to the actual fish. The flow rate can be regulated with manual or automated controls. Only consider this option in smaller temporary systems, such as a purge or a quarantine system.

Each oxygenation method has its advantages and disadvantages and the best solution may actually be a combination of oxygenation methods. Common combinations include the use of LHOs with oxygen cones, or oxygen cones with fine bubble diffusers. In the first case, LHOs handle a majority of the oxygenation needs in a farm, but when there are spikes in oxygen demand, side stream pumps can pressurize the oxygen cones and deliver extra oxygen to the system. The second combination is a similar scenario where the oxygen cones supply a bulk of the oxygen flow, while the fine bubble diffusers automatically turn on during high oxygen demand times, such as right after feeding. Using two methods in tandem gives the operator more flexibility while still keeping energy costs low and overall dissolution efficiencies high. Whatever option you choose make sure you can continuously meet the necessary oxygen demands of the fish.

Oxygen can be supplied to the oxygenators via one of two methods, an oxygen generator, or liquid oxygen (LOX) tanks. In the second case LOX tanks are placed onsite at the farm and refilled regularly by a LOX vendor. The oxygen generator, in contrast, uses air to make a concentrated oxygen flow (>90%) on site. Both systems are discussed in more detail in the following chapter. Each needs to be appropriately sized for the oxygen needs of the farm, the LOX tank needs to have

Chapter 5: Water Treatment Components

adequate volume and the oxygen generator needs to be able to supply the appropriate amount of oxygen in liters per minute. The two methods each have their advantages; the capital costs of LOX system are lower than then the cost of an oxygen generator. However, when using LOX you must pay for the oxygen from the vendor typically on a per liter basis. The operating costs of an oxygen generator are mostly the electricity needed to operate the generator. In most RAS where electricity rates are reasonable, an oxygen generator is the correct choice to save money in the long term. In both cases though it is important to ensure there is both adequate flow and pressure to supply oxygen to all of the oxygenators. This is mostly a concern for oxygen generators where the output pressure may be as low as 1 bar. This may only be enough pressure for low head oxygenation methods. As a result, the oxygenation method you choose may also determine what oxygen supply equipment is needed.

The use of pure oxygen requires special considerations because of its flammability and corrosivity. All equipment associated with the oxygenators needs to be corrosion resistant; this can include fiberglass, some plastics and rubbers, along with high-grade aluminum and stainless steel. Similarly, all equipment needs to be spark-safe to reduce the chances of fires or explosions. An oxygen resistant rubber tube should always be used to connect the oxygen supply with the oxygenators. Make sure the tube diameter is large enough to handle the necessary flow rates required. Keep these safety concerns in mind when purchasing and installing your oxygenation system.

Chapter 5: Water Treatment Components

Oxygenators require continuous monitoring and maintenance since a dip in dissolved oxygen levels can lead to sudden fish mortalities. The best oxygenator will dissolve oxygen into the water with minimal energy costs and high oxygen dissolution efficiency. Capital costs vary between the various methods and the correct choice will depend largely on the water temperature and salinity, along with the stocking density and feed rate. High temperature waters with large oxygen demands will require more energy intensive methods of oxygenation, while lower stocking densities in cool water systems can use low-head options.

Pumping

Pumps are the real workhorses of any farm, providing either the necessary water pressure or raising the water high enough so it can flow through all of the water treatment steps. This also means that pumps use a majority of the electrical power in any RAS. The two major factors that determine the power usage of the pumps is the water flow rate and the head pressure. The main water treatment water flow rate is calculated by multiplying the total system water volume by the chosen HRT (see System Volume Calculation section). Side stream water flow rates are dependent on the treatment step and the sizing of the equipment. As a result, water flow rates are relatively fixed and decreasing the flow rates in the name of energy conservation will lead to compromised water quality.

Chapter 5: Water Treatment Components

On the other hand, the pumping head is determined by the design of the farm and the water treatment methods that you choose. The challenge of designing an energy efficient farm is the challenge of lowering the total head needed for all the water treatment steps. The hydraulic head of a pump is equal to the theoretical height it can raise water. In addition to raising the water up a number of other factors determine the total hydraulic head of a water treatment system. Water pressure, water velocity, and friction head losses all increase the hydraulic head and can all be quantified in terms of meters of water as shown below in Bernoulli's equation.

$$H_T = (z_2 - z_1) + \left(\frac{P_2}{\rho g} - \frac{P_1}{\rho g}\right) + \left(\frac{V_2^2}{2g} - \frac{V_1^2}{2g}\right) + h_l$$

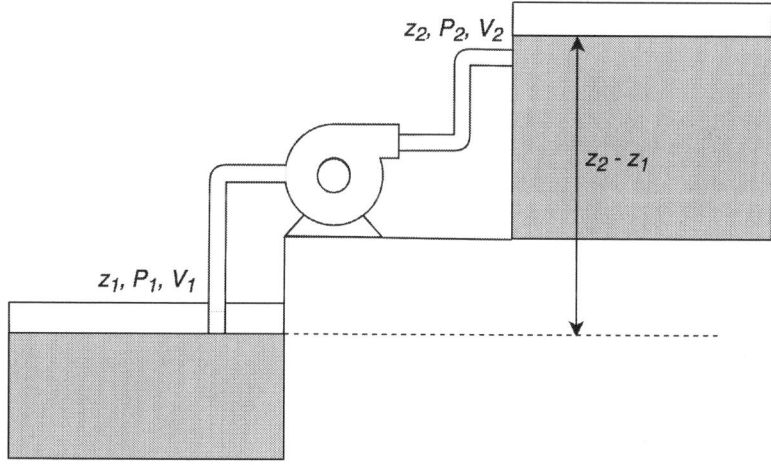

Figure 12. Bernoulli's equation inputs include water height, water pressure, water velocity and friction head losses.

Chapter 5: Water Treatment Components

Where H_T is equal to the total dynamic head, z_1 and z_2 are equal to the water height, P_1 and P_2 are equal to the pressure, ρ is equal to density of the fluid, g is equal to the gravitational acceleration, V_2 and V_1 are equal to the water velocity, and h_l is equal to the sum of frictional head losses. Frictional head loss is increased by narrow pipes, valves, or pipe constrictions and can be calculated for any given system by using pipe hydraulic equations. By summing everything together Bernoulli's equation allows us to determine the total pumping head needed for any system. Most of the factors that go into determining the total pumping head can be controlled through the design of the farm. And as shown in the equation below, increasing the head will increase the power usage and size of the pumps.

$$P = \frac{\rho g Q H_T}{\eta}$$

Where P is equal to the pumping power required in watts, ρ is equal to the density of the fluid (ρ= 1000 kg/m³ for freshwater), g is equal to the gravitational acceleration (g= 9.8 m/s² on Earth), Q is equal to the water flow rate in cubic meters per second, H_T is equal to the total head in meters, and η is equal to mechanical efficiency of the given pump, usually between 0.6 and 0.85.

The mechanical efficiency of the pump is determined by both the type and size of pump for a given flow rate and total head. We can identify the correct pump for any job once the total head and water flow rate are known. This is done through the

Chapter 5: Water Treatment Components

use of pump curves; these curves have water flow rate on the x-axis and head on the y-axis. A parabolic shaped curve shows the maximum head the pump can achieve for any given flow rate. Frequently there is also an additional contour map below the curve that shows the mechanical efficiencies of the pump at all of the possible head and flow rate points that can be achieved by that pump. Often times the same pump will be available with a variety of motor sizes, and each motor size will have its own curve. Pump manufacturers create pump curves through standardized experimental testing, and often they have the ability to generate custom curves for your specific application. Finding the most efficient pump for any given set point requires knowledge of various pump manufacturers and involves reviewing a lot of pump curves. Often times it can be frustratingly difficult to find an efficient and cost-effective pump for every set point in your RAS. Pumps have the unfortunate drawback of varying in price by an order of magnitude between manufacturers for what is essentially the same pump. As a result a lot of trial and error is needed to find the correct pump for each application.

Rarely does the flow rate or total head remain constant for any main or side stream water flow. Changing the flow rate will change the maximum head the pump can achieve and the mechanical efficiency. It is important to first properly size the pump so that it can handle the maximum flow rates and total head since lower flow rates and heads will also be achievable by that same pump. The easiest way to maintain high mechanical efficiencies with changing flow rates is through the use of variable speed pumps. Pumps that are powered by

Chapter 5: Water Treatment Components

alternating current (AC) power can have their impellers slowed down or sped up by changing the electrical frequency. Varying the speed of the impeller affects the pump's power usage and the maximum head at a given flow rate that the pump can achieve. Specially made variable speed pumps with built in controls can be purchased, but it is more common for standard pumps to be electrically wired to a variable frequency drive (VFD), which can alter the standard electrical frequency (typically 50 Hz or 60 Hz). VFDs are commonly used with all styles and sizes of pumps across a variety of industries and they are also commonly used on fans and blowers. Any pump with a 2 Kw or greater motor is best controlled by a VFD. A VFD can be manually controlled, or when used in tandem with the appropriate sensors, it can automatically adjust the speed of the pump to maintain a constant pressure or flow rate. VFDs have the added benefit of being able to slowly ramp up and ramp down the pumps speed, preventing damage that may occur during sudden starts or stops that would occur with out a VFD. All pumps should be wired with some amount of electrical controls and alarms. That way the pump can automatically shut off and alert the operators in case of overheating, excessive current, or any number of other issues that could occur.

In addition to knowing the pumping head and water flow rate there are a few other factors to consider when selecting a pump. First, you will want to also know the net positive suction head of the pump (NPSH), this is equal to the height the water must be raised before it enters the pump, plus friction head losses on the suction side of the pump. The NPSH can never surpass 10.3 meters since it will then be greater than

Chapter 5: Water Treatment Components

atmospheric pressure, which will lead to cavitation inside the pump. Cavitation is essentially the water boiling under near vacuum conditions. Cavitation will lead to the pump making loud noises and it will quickly destroy the pump by introducing gas into the pump body.

Next you will need to determine how the pump will be primed whenever it is first started and any time after that. For non-submerged pumps the pump body typically needs to be full of water before it will function properly and suck new water in while pushing water out. Priming does not need to take place if the pump is located below the water's surface on the suction side of the pump. It will be needed if there is any NPSH. Pumps can be primed through the use of a hose attached to a specially designed port that will flood the pump before it is started. Or it is also common to have a small foot powered pump that can raise water and fill the pump before it is started. Self-priming pumps are also common and typically have a priming pot attached to the suction side of the pump that remains full of water even when the pump is shut off and the water drains out of the rest of the pipes. All methods work well but at least one needs to be chosen for all pumps that require priming.

Pump motors are typically available in a few different styles. All aquaculture pumps should include totally enclosed fan cooled (TEFC) motors. These motors are better suited for high moisture environments where water can easily get inside the motor and hasten the corrosion of the electrical windings. Open drip proof (ODP) motors are the other frequently seen

Chapter 5: Water Treatment Components

option, but these should be avoided since they do not have a long lifespan in most aquaculture environments. In addition to the motor type, the motor also needs to be designed for the correct electrical voltage, frequency, and phase. Voltages vary between 120V to 575V depending on what country you are in and what kind of power service is available at your site. Typically larger motors are best operated at higher voltages to decrease the current and raise the efficiency. The standard electrical frequency is either 50 or 60 hertz (Hz) depending on what country you are in, 50 Hz pumps will spin more slowly and deliver less pumping power than a 60 Hz motor. The electrical phase is either single or three phase, only three phase power can be used in conjunction with a VFD and it should always be used if available at your site. The motor will also have a power rating in either horsepower (HP) or kilowatts (Kw) that designates the maximum power the pump will draw. It is important to realize that most pumps rarely use the amount of power that the motor rating specifies. Instead the power usage will depend on the load that the pump must handle. Typically higher heads will lead to higher power usage. Nonetheless, the VFD and all other electrical components need to be sized correctly for the amount of maximum power that the pump motor can draw. Finally, the motor will have a standard frame size since electrical motors are not made by the pump manufacturer but instead by a motor manufacturer. The motors standard dimensions allow it to be easily specified and replaced in case of failure. All motors are not created equal and the correct motor will allow you to use the pump in an aquaculture environment with the power that you have available at your farm site.

Chapter 5: Water Treatment Components

A final consideration for pumps is the materials used to construct them. Corrosion resistant pumps are typically constructed of 316 SS, plastic, or titanium. Less corrosion resistant materials include 304 SS or coated cast iron. It is also common to see a variety of materials used in a single pump with the impeller typically having the more corrosion resistant material. Material choices can lead to huge discrepancies in price between the same pump. 316SS and titanium pumps would generally be considered expensive but are needed in warm saltwater environments. Plastic pumps are typically available in sizes up to 4 Kw and are an inexpensive alternative for use in corrosive environments. Cast iron pumps are often the most cost effective and work well for almost all freshwater applications. When selecting a pump understand what materials are being used in the pump body and impeller, and choose the materials that are best suited for your farm.

All pumps need to be monitored and regularly maintained. Regular maintenance includes greasing bearings and cleaning the fans. Fortunately pumps are fairly robust, simple pieces of equipment and all parts can be easily replaced if they wear out. Impellers may need to be replaced periodically if they or their bearings wear out. Low speed pumps reduce the wear on the impeller, but still require eventual replacement. Motors can be easily and inexpensively replaced on non-submerged pumps. Submerged pumps have more specialized motors that can still be replaced if needed. It is important to have replacement parts or entire pumps on hand in case of pump failure. Pumps are crucial to all of the water treatment processes and the flow

Chapter 5: Water Treatment Components

rate needs to be maintained at all times. As a result, quickly fixing or replacing pumps should be a priority. This may mean sourcing pumps from local suppliers who can quickly send replacement parts.

Multiple identical pumps should handle the main treatment flow for all RAS. This provides a degree of safety since a single pump failure will only result in a partial loss of water flow. Ideally, three to four pumps should be responsible for the main treatment flow in each system. Identical pumps also make stocking spare parts straightforward. In addition, pumps should be located as close to the suction point as possible to reduce NPSH. Similarly, they should be oriented and placed to reduce turns and distance in the pipe between the pump and the outlet. Frictional head losses should be a consideration with the placement of all equipment, and placing the pumps near the center of the system reduces the distance needed to travel to each tank. Large, heavy pumps should also be located where they can be easily serviced and replaced if necessary. This may include the installation of an overhead lift or space for a portable lift to be moved into place when needed. All of these considerations make the pumping system more robust, efficient, and easier to maintain.

There are a nearly endless number of designs and styles of pumps, each best suited for a particular application. However, there are a few styles that are predominately used in aquaculture systems. These include submerged axial flow pumps, line-shaft pumps, submerged and dry centrifugal pumps, and booster/pressure pumps. Each style of pump is

Chapter 5: Water Treatment Components

best suited for a specific application within a RAS. Typically a particular pump style will have a sweet spot of heads that it works best for, with lower head pumps usually being best suited for larger flow rates and vise versa for high head pumps.

Axial flow pumps are an entire class of pumps that includes both submerged axial flow pumps and line-shaft pumps. All axial flow pumps employ essentially the same technique to move water. A propeller is placed in line with a large pipe, and the propeller spins at a high speed to move water along the pipe. Axial flow pumps are ideal for low head, high flow applications and are commonly used for the main water treatment flow in large scale RAS. Submerged axial flow pumps are designed such that the entire pump, including the motor, are submerged underwater. The pump rests on the bottom of a tank or sump and is arranged in a vertical configuration where water is drawn off the bottom of the tank up through the pump body and ejected out the top into a pipe that extends above the pump towards the water's surface. The maximum head achievable by this style of pump is rarely over 7 meters. The sweet spot is in the 3-5 meter range. This style pumps is able to supply a huge range of flow rates and is typically available in sizes ranging from 1 Kw to truly huge 180 Kw pumps, however the maximum size in large aquaculture systems is closer to 30 Kw. The pumps are cooled by the water flowing over the pump body as opposed to the use of fans, which are used on all non-submerged pumps. One advantage of this is the pumps do not create any audible noise in the farm. Waterproof, heavy-duty electrical cables supply power to the pump. The design of the submerged axial flow pumps is simple and largely maintenance

Chapter 5: Water Treatment Components

free. Over time though, the mechanical seal may fail, allowing water into the motor. The drawback of any submersible pump is that when it fails it needs to be detached from the outlet pipe on top of the pump and hauled up to the water's surface for repair or replacement.

An alternative axial flow pump design is the line-shaft pump. In a line shaft pump the propeller is located near the bottom of tank and is attached to a shaft that extends upwards through a pipe to above the water's surface where it connects to the electrical motor. The electrical motor is mounted vertically to directly spin the propeller on the shaft below it. The water moves up the pipe past the propeller and typically makes a 90-degree sweeping turn once it rises above the water's surface. Same as the submerged axial flow pump the optimal heads for this style of pump are rarely over 7 meters and it is best suited for heads of 3 to 5 meters. Any imaginable aquaculture appropriate flow rate can be achieved with a line shaft pump. Line shaft pumps operate under less harsh conditions since they are not under water and they can be repaired more easily since they are readily accessible at all times. The disadvantages are that the pumps can be more expensive than submersible pumps, largely due to the expense of the submerged line shaft casing, and a more complex design. Either axial flow pump option is the best choice for low-head high-flow applications.

Centrifugal pumps, also known as radial-flow pumps, are the most common style of pump across all industries including aquaculture. They are well suited for a wide range of heads and flow rates. However, The minimum efficient head for most

Chapter 5: Water Treatment Components

centrifugal pumps is around 5 meters. In centrifugal pumps, the water enters at the rotational axis of the impeller where it is then spun outwards against the pump body; next the water is essentially slung out of the pump body into the outlet pipe. The pump uses centrifugal inertia to move the water, thus its name. Submerged centrifugal pumps, also commonly called sump pumps or trash pumps, are set up similarly to submerged axial flow pumps. The pump is placed on the bottom of a tank or sump and water is sucked up off the bottom into the pump and sent upwards to the water's surface through an outlet pipe. Heads of 15 to 40 meters are common with this style of pump in power ratings of 2 Kw to 20 Kw. Same as other submerged pumps, they are cooled by the water passing over them, make very little audible noise, and can be difficult to service. Submerged centrifugal pumps are appropriate for side stream flows that require a higher head than is achievable with axial flow pumps.

Non-submerged centrifugal pumps, also just called centrifugal pumps, are more common and more cost-effective for most applications. They are best for heads of 5 to 40 meters and can handle any flow rate. Centrifugal pumps are a great option for all side stream flows with heads of 5 meters or greater. Most centrifugal pumps can be coupled with a priming pot or check valve to make priming automatic. The disadvantage of the centrifugal pumps is they need to be mounted properly to avoid issues with vibration and some designs are noisy. The advantage is centrifugal pumps are manufactured in large quantities since they are used in a variety of industries, making their price relatively low compared to many other pump styles.

Chapter 5: Water Treatment Components

Plastic swimming pool centrifugal pumps are particularly inexpensive, but often times not as reliable as other styles. For all smaller centrifugal pumps it is a good idea to have complete spare pumps on hand in case of failure.

Booster or pressure pumps are used almost exclusively in conjunction with a drum filter. Booster pumps use multiple impellers to deliver a small flow rate of water at an extremely high head. The head may be as high as 100 meters and often is measured in bars instead of meters. Three bars is a common pressure specified for drum filter cleaning, and the flow rate will depend on the size of drum filter, but it is rarely greater than 5 m^3/hr. Typical power ratings are 0.5 to 3 Kw, again depending on the size of the drum filter. The pumps are typically mounted near the drum filter and come in a vertical configuration with the motor sitting on top of the pump body. A single booster pump can be used for multiple drum filters, but it is better to have replicate pumps. This can mean either one dedicated pump per drum filter or the use of a pressure pump skid that uses multiple pumps to supply a constant pressure of water to multiple drum filters. In either case a single pump failure will not result in the entire solids filtration system going down. Many booster pumps come with specialty flanges on both the inlet and outlet. Ideally the pump can be supplied with threaded ports instead since this make tying the pump into the system significantly less expensive. For aquaculture purposes, booster pump pressures are not so high that they require special high-pressure piping.

Chapter 5: Water Treatment Components

To summarize, in order to select the proper pump you will need to have following information on hand: water flow rate, head, NPSH, available electrical voltage, phase, and frequency, along with the desired pump material, motor type, and priming method. The water flow rate and head will narrow the choices for which pump style is appropriate, the price and pump efficiency will then determine which exact pump style and size you end up selecting. Pump efficiency is particularly important since it can lead to significant energy savings over the course of the pump's lifespan. Similarly, with an efficiently designed farm, the head should be minimized to minimize the pump's energy usage. Work with an experienced engineer or multiple pump distributors to help identify the best pump selection for your operation. If a pump is not operating reliability or efficiently, don't be afraid to switch that pump out for one from different manufacturer. The perfect pump for each application is impossible to find, but the best pump can be found with some work and the proper preparation.

Fine Particle Filtration

Fine particles are defined as suspended and dissolved solids less than 30 microns in diameter. Given their small sizes the particles are difficult to remove with traditional solids filtration methods. Nonetheless their removal from the system's water leads to clearer water and higher water quality. Fine particles can irritate fish's gills, increase biological oxygen demand, and make the water an off-putting brown color. The two widely used methods for removing these particles is foam fractionation

Chapter 5: Water Treatment Components

and ozone treatment, and often both are used in tandem for maximum efficiency.

Foam fractionation is also known as protein skimming and it involves bubbling a gas through the water that leads to the creation of foam at the water's surface that can then be removed. This foam is made up of small particles, proteins, fats, and other organic substances that interact with the surface tension at the water-to-gas interface. The bubbles create the environment for these suspended and colloidal solids to collect into a brown foam at the water's surface. Substances that can be removed with foam fractionation are typically hydrophobic and are known as being surface active or surfactants, which means they interact with the bubble's surface. A number of factors influence how much foam is formed and what types of particles it removes. For example, foam fractionators work better in saltwater systems and worst in hard water systems. The amount of foam will also vary with the amount of feed added to the water with a lot of foam forming in the thirty minutes following a feeding event.

There are a number of different designs for foam fractionators but they all perform the same basic function (Figure 13.). A gas, either air or ozone, is bubbled into a vessel using either airstones or a venturi. Water enters and exits the vessel below the surface level. At the surface, foam may overflow from a weir into a collection vessel, or there may be an actual mechanical skimmer that regularly pushes the foam into a collection vessel. The collection vessel drains the foam away to waste as the bubbles pop and it turns into a thick brown liquid. Some foam

Chapter 5: Water Treatment Components

fractionators include automated spray nozzles that wash foam away regularly. Foam fractionators can look like a complex system of pipes and tubes, but their actual operation is reliable and they are easy to maintain.

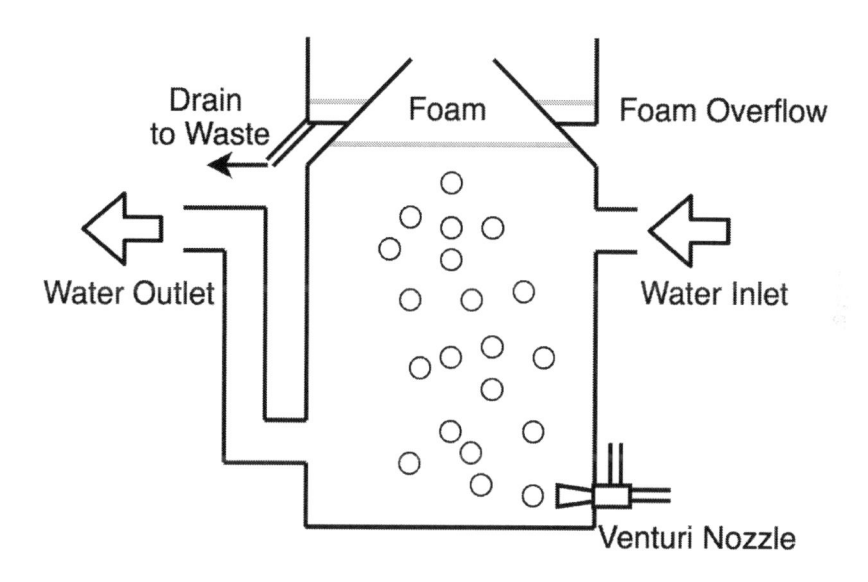

Figure 13. Foam fractionator cross section

Foam fractionators are typically operated on a side stream and the flow rate determines the size of fractionator that is used. Typically the foam fractionator will be sized for the inflowing water to have a contact of between 90 to 120 seconds inside of the vessel. Determining the amount of water that should be sent through a foam fractionator is debatable, but aim for around 5% of the main flow. Equipment manufacturers can help with the appropriate sizing of a foam fractionator based on the amount of daily feed in the farm. Venturi aerators are commonly used to create the bubbles in foam fractionators, but airstones can also be employed. Smaller bubbles are more

Chapter 5: Water Treatment Components

efficient at removing solids since they create more surface area for the same amount of airflow.

Ozone can be used alone or in tandem with a foam fractionator. As mentioned before in the disinfection section, ozone can also help breakdown organic compounds and flocculate fine solids. Flocculation is the process of small particles being attracted to each other to form larger particles that can then be removed in the solids filtration step. Ozone also oxidizes dissolved proteins, fats, and other organic substances, typically making them more hydrophobic and more likely to become surfactants that can be removed in foam fractionation. This also has the affect of removing some of the brown water color that is found in many RAS. Significantly less ozone is needed for water quality maintenance than is needed for disinfection. A dosage of 10-20 grams of ozone per kilogram of feed has been found to be effective in increasing the water quality without leaving an excessive residual concentration in the water. Ozone can be bubbled into a contactor similar to what was described in the disinfection section or it can be directly injected into the foam fractionator where it will mix with the air. Another option is to inject the ozone into the oxygen cone along with the oxygen. Ozone is particularly useful in freshwater systems where foam fractionation is more difficult. The same safety concerns as described in the disinfection section for fish and humans must be observed. Also residual ozone needs to be removed from the water before it is returned to the fish tanks, either by consumption, off gassing, or ultraviolet radiation.

Chapter 5: Water Treatment Components

Many farmers initially ignore the importance of fine particle removal since it is not always emphasized as a critical filtration step. However, many of these farmers later realize that the higher water quality is worth the additional filtration step and they end up adding a jerry-rigged system after the fact to help remove fine particles. It is better to instead plan from the beginning that fine particle removal will be needed. Foam fractionators should be strongly considered in all saltwater systems where they operate more effectively. In freshwater systems they are most likely not needed and instead only ozone should be relied on to remove fine particles. The disadvantages of foam fractionators are they can be large, expensive and require an additional side stream pump. The top of the unit should be cleaned regularly to prevent the build of brown sludge and ensure the waste is flowing out of the unit. The addition of ozone requires operating an ozone generator and monitoring the redox potential to make sure residual ozone is removed before reaching the fish. However, the increased water quality that is achieved from both of these methods is worth the investment and operating expense.

Heating & Cooling

Temperature regulation of RAS is key to providing the proper growing environment for the fish since it determines the rate of many biological and chemical reactions. Depending on the local climate conditions and the species selected either heating or cooling may be required throughout the year, and in some cases both may be required. Water has a relatively high heat capacity of 4.18 J/(g°C), this means that a relatively large

Chapter 5: Water Treatment Components

amount of energy is needed to change the water's temperature. This is good for maintaining temperature stability, but bad when a lot of heating or cooling needs to take place to reach an optimal temperature.

The exact heating or cooling needs of the farm can be difficult to estimate. These loads are often measured in watts or British thermal units (BTU), the later is often used with heat pump equipment. For incoming water the heating/cooling load is easy to calculate if you know the incoming average water temperature (T_{in}) and the system water temperature (T_{system}). The heat load will be equal to the temperature difference multiplied by the heat capacity of water (C), water density (ρ), and water flow rate (Q). This will give the heating/cooling load in watts.

$$Heat\ Load = \rho Q C \left(T_{system} - T_{in} \right)$$

Calculating the load to maintain the system water temperature is more difficult and will depend on local weather conditions including outdoor temperature, humidity, and wind speed. It will also depend on the farm building's insulation, air exchange rate, and total system volume. Evaporative heat losses are particularly difficult to estimate, as is the heat generated by pumps and other equipment. The operation of equipment such as pumps and blowers creates a lot of waste heat that ends up either in the water or in the air. This waste heat is not so wasteful when heating is needed, but it can lead to higher energy costs when water-cooling is needed. Hire an experienced engineer to help you calculate these loads or experiment yourself to figure out the heating and cooling

Chapter 5: Water Treatment Components

needs of your farm. Many farms operate undersized equipment at near maximum loads during the summer or winter. Avoid doing this and get the right equipment that can easily handle your loads.

There are two common methods for changing temperature in aquaculture systems, the first method is direct heating and the second is heat pumping. Direct heating is typically done with either electric resistivity or the burning of natural gas and can only raise the water temperature. Heat pumping on the other hand can both lower or raise the temperature of water. A heat pump works by moving heat from a heat source to a heat sink, usually by compressing and expanding a working fluid such as Freon (Figure 14). Moving heat rather than creating heat is more efficient and often times cheaper. A heat pump can often move three to four times as much energy in the form of heat than the electrical energy needed to operate the heat pump, this is known as the coefficient of performance (COP). Heat pumps are the only option if water needs to be cooled; both air conditioners and refrigerators are examples of heat pumps. Some heat pumps are reversible meaning they can be operated to both heat and cool the water. The disadvantage of a heat pump over a direct heater is the initial capital costs for a heat pump are much higher and there are more moving parts that require maintenance. Nonetheless in most scenarios choosing a heat pump will save money over the lifetime of the equipment due to its lower operating costs.

Chapter 5: Water Treatment Components

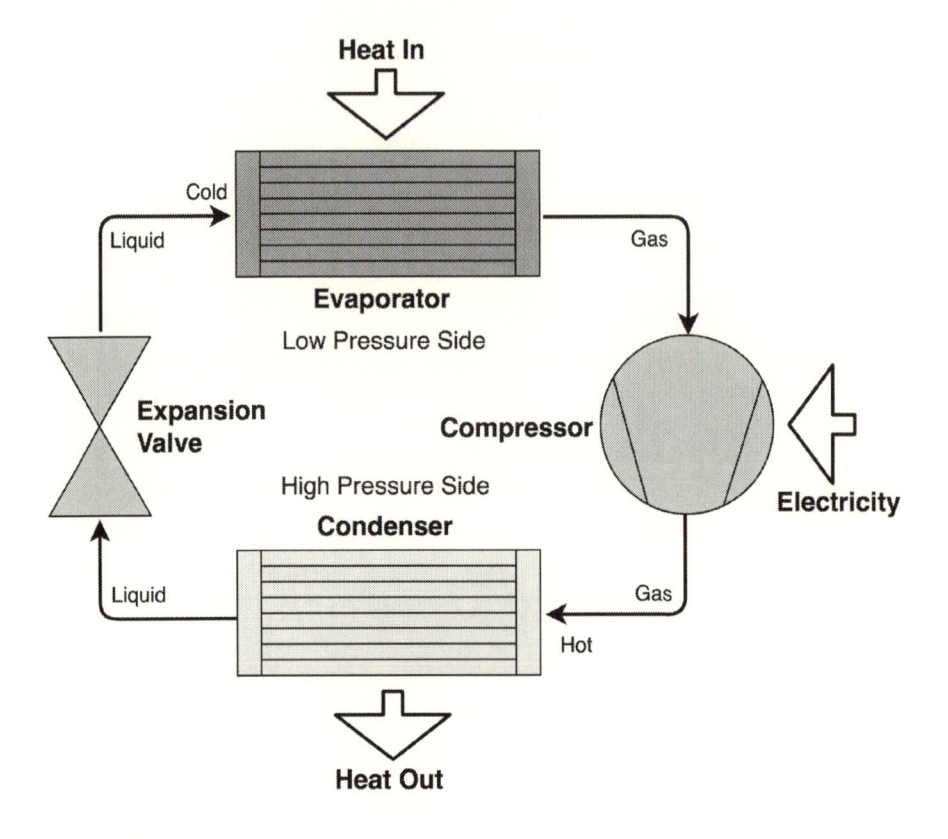

Figure 14. Heat pump flow chart.

Temperature regulating equipment can either be operated on a side loop independent of the main water flow, or heat exchangers can be submerged directly in the main flow, often times in the pump sump. The equipment will dictate which method is used. Temperature is easy to monitor and control. A thermocouple can accurately measure the temperature within 0.1 °C, and just like in a house, a thermostat can turn heating and cooling equipment off and on to maintain the systems temperature within a narrow band. Each system can be

Chapter 5: Water Treatment Components

operated on a separate thermostat so that the temperature can be independently operated and shut off if needed.

Heat exchangers can be used to either cool or heat the water and come in a variety of shapes and designs. At its simplest, a heat exchanger is a vessel that has high surface area where a hot or cold working fluid, sometimes water or glycol, is pumped through a series of metal tubes or plates that are exposed to the system water. The high surface area, high thermal conductivity of the metal, and the large temperature difference between the working fluid and system water all facilitate heat transfer. To maximize heat transfer the working fluid and system water should flow counter-current to each other.

There are two major types of heat exchangers: Shell and tube heat exchangers and plate exchangers. The shell and tube exchangers are often used with chillers and consist of an outer shell that the system water flows through that is filled with a series of narrow tubes that the working fluid flows through. In a plate exchanger the working fluid flows through thin plates that are either directly submerged in the system water or are inside of a vessel that system water is pumped through. Because of their high surface area, all heat exchangers require some cleaning since biofouling will reduce the heat transfer rate, and in the case of plate heat exchangers biofouling could prevent water flow. For maximum corrosion resistance titanium should be used in heat exchangers, followed by stainless steel. In order to properly size a heat exchanger, the temperature difference between working fluid and system water must be

Chapter 5: Water Treatment Components

known along with their respective flow rates. This information, along with the thermal conductivity of the material of which the heat exchanger is constructed, gives enough information to determine a surface area. For some reason equipment manufacturers are coy on the actual surface area of their heat exchangers, but they are often willing to help you select which product in their range is correct for your operation if you first give them the appropriate data for them to do the calculation.

Heating can be accomplished with a variety of equipment. Submerged electrical heaters are inexpensive and easy to operate, they are available in sizes from 100 to 6000 watts and they often require a 230V or 460V electrical connection. The heaters are available in a variety of geometric configurations, such as vertical or horizontal tubes, and various material choices, such as stainless steel and titanium. Most electrical heaters include overheating protection that will cause a breaker to trip if the heater becomes too hot. A resettable breaker option is a must have for this kind of heater since they can often overheat if the water levels become too low and the heater is no longer submerged.

Another option is an inline electrical heater; the temperature change across the heater is not great, but they can be sized to deliver up to 20,000 watts of heating. For even larger heaters an inline natural gas heater is a good option; they can be sized up to 100,000 watts of heating. Natural gas boilers can also be used in tandem with heat exchangers to heat a working fluid that is recirculated through the heat exchanger in order to warm the water. A final option for heating is the use of outdoor

Chapter 5: Water Treatment Components

mounted solar heaters. The heating loads for these will depend on the size of the equipment, the farms location and seasonal weather changes. Solar heaters can also be used in tandem with heat exchangers to eliminate the need to pump system water through a system that can easily biofoul. All heaters are subject to biofouling, and in locations with hard water, calcium or magnesium scaling can occur on the heater. The final method is to use a heat pump, however the minimum sized unit is around 300 watts. A heat pump is a good option for water temperature maintenance and needs to be used in tandem with some type of heat exchanger.

All of these methods have their advantages and disadvantages. As discussed before, the heat pump will use less energy for the same heating load, but has higher capital costs. Electrical heaters on the other hand are easy to install and operate; however natural gas is often even cheaper than electricity. Natural gas heaters are reliable and the best option for heating water that is coming into the farm for the first time. For system wide temperature maintenance, natural gas heaters are best operated with a boiler and heat exchanger plates. The correct equipment choice will depend on your heat loads and the cost and availability of electricity and natural gas at your location.

Water cooling is done exclusively with a heat pump, often called a chiller when it is being used to cool the water. Chillers can be either air or water-cooled. This means that either air or water is moved across the condenser to pull heat out of the heat pump. In both types of chillers the system water moves across the evaporator to pull heat out of the system water. In

Chapter 5: Water Treatment Components

this case the evaporator is commonly called a chiller barrel and is actually a shell and tube heat exchanger that system water is pumped through. The working fluid is the freon in the chiller. Chillers are often designed to only cool a side stream of water and they operate best in a temperature range of between 4°C to 26 °C.

Air cooled chillers look similar to traditional split air conditioner units used in many homes. The condenser fans take up considerable amount of room and they need to be located out of direct sunlight where they can receive fresh air. In a water-cooled chiller a small stream of cool water is used to pull heat out. These chillers are smaller but the water that is used often times is sent down the drain. The efficiency of the chiller will depend on the temperature of air or water used to cool the condenser.

Not only can equipment be used to heat and cool the water but heat can also be conserved in the system with insulation, heat exchangers, and good practices. Conservation techniques reduce energy costs and heating or cooling equipment costs. Firstly, all RAS should be operated inside of an insulated building. This reduces temperature swings and energy costs due to seasonal and daily weather changes. Next the air inside of the farm should be kept at a temperature near the water temperature. In cool water systems this typically means the air temperature is a few degrees above the water temperature, and in warm water systems the air is usually only slightly cooler than the water temperature. The air temperature can be maintained with HVAC equipment that will be discussed in a

Chapter 5: Water Treatment Components

following section or by reducing the exchange of air to the minimum needed to keep carbon dioxide concentrations within reasonable limits. Another method is to ventilate the farm each day only when the outside temperature is most similar to the desired indoor temperature. Heat can also be conserved through the use of a heat exchanger on the wastewater leaving the farm. All incoming water and wastewater can be sent through a heat exchanger whereby the incoming water gains some of the heat being sent out of the farm that is in the warm wastewater. This can also be used to cool the incoming water if needed. A final consideration to reduce heating and cooling energy costs is to locate the farm in a suitable location where the seasonal outdoor temperatures are similar to the required system water temperature.

Heating and cooling are potentially big energy users for any farmer and can be difficult to estimate. The correct equipment selection and sizing will reduce capital and operating costs. Careful consideration should also be given to how heat can be conserved within the system. Increased capital costs are worth the investment to decrease energy needs and their associated operating costs.

Alkalinity & pH Control

pH plays an important role in a number of biological and chemical processes in RAS. For example, at pHs of 6.8 or lower the nitrification rate of the biofilter begins decreasing. Similarly, we discussed in the carbon dioxide stripping section how lower pHs can negatively affect fish health by changing the fish's

Chapter 5: Water Treatment Components

blood chemistry. pH also affects the toxicity of various metals and the toxicity of TAN. All of these reasons make pH a critical parameter to monitor and adjust.

Alkalinity and pH are generally managed in tandem primarily with the goal of lowering the pH and raising the alkalinity to levels above 50 mg/L, this helps to manage pH swings. The biofiltration process lowers the pH of the system during the first step of nitrification when hydrogen ions are produced by the oxidation of ammonium. These hydrogen items shift the pH downwards and consume alkalinity. In order to keep the pH of the system near a constant we have to regularly add alkalinity to the system. Similarly, dissolved carbon dioxide, as carbonic acid, shifts the carbonate system towards lower pHs (see Figure 7). Adding carbonate and bicarbonate to the system can shift the carbonate system back towards a neutral pH. It is uncommon that the pH needs to be adjusted downward, but this is also possible with the use acids.

Alkalinity can be added to an aquaculture system in a number of different forms, most commonly hydroxides and carbonates are used. Common carbonates include sodium bicarbonate ($NaHCO_3$), commonly known as baking soda, and calcium carbonate ($CaCO_3$), also known as lime. Common hydroxides include sodium hydroxide (NaOH), known as lye, and calcium hydroxide ($Ca(OH)_2$), known as slaked lime. Both should be handled with care, as they are caustic substances. Carbonates are recommended for their price, and ease of storage, the downside is they take longer to dissolve. Whenever adding

Chapter 5: Water Treatment Components

alkalinity to the system it is important not to overdose, especially with hydroxides, which can rapidly change the pH.

Alkalinity is added to the system after the carbon dioxide stripping occurs, typically in the pumping well before water is returned to the fish tanks. This allows the alkalinity to be well mixed into the water before being exposed to the fish, and by being applied after the carbon dioxide stripping, it allows the stripping process to be more efficient. There are a few methods for mixing in carbonates, the simplest is to monitor pH manually with an electronic sensor and toss in scoops of lime or baking soda into an area of the system where it will mix well and dissolve quickly. No equipment providers offer a plug and play automated device at the moment, but one can be easily constructed. One easy method is to create a lime slurry tank with a 400 liter open top tank. Aeration can provide mixing to keep the lime in suspension. A dosing pump can then add the lime slurry to the tank with feedback from a pH sensor. In smaller systems a bed of lime or oyster shells can be put inline with the system, where water passes over or through the bed, slowly adding alkalinity to the system. This can often only be used on a small scale, but it can be a good option for nursery systems.

Chapter 6: Additional Equipment Considerations

Waste Treatment

Aquaculture waste treatment deserves an entire book of its own, but it is good to have an idea of what may be needed for the average RAS. Waste can come in the form of wastewater, waste solids, or a combination of both. Each category can be treated with different techniques, many of which are borrowed from wastewater treatment for human and animal sewage. Fortunately, by sewage treatment standards aquaculture wastes are relatively benign and easy to treat. The goal of all waste treatment is to clean the water or solids so that they are devoid of environmental pollutants. A bonus of waste treatment is that it can yield useful byproducts that can recycle nutrients back into other agriculture activities.

The degree of waste treatment will depend largely on the local discharge regulations. In the US a national pollution discharge elimination system permit (NPDES) is needed for any water that is discharged. The permit outlines the acceptable parameters of the wastewater before it is sent out of the farm and back into the natural environment. Beyond the local regulations it is also a good idea for any RAS operator to do their part in protecting the environment. After all, one of the goals or RAS is lowering the environmental footprint of fish farming and this includes treating or recycling any waste streams.

Chapter 6: Additional Equipment Considerations

The volume of wastewater is approximately equal to the water that is replaced each day in the system. The replacement rate may be between 1 to 10% of the total system volume per day, less water replacement makes wastewater treatment easier. The first step in wastewater treatment is removing any solids, usually this is done with a radial flow separator or settling basin since the water volumes are relatively small and the waste treatment system can be located outdoors where there is more room. After all of solids have been removed from the wastewater the next step is denitrification and phosphorous removal.

Denitrification of the water can be accomplished through a couple of different biological methods. The first method is through the use of an anoxic biofilter. In the absences of oxygen anaerobic denitrifying bacteria will reduce nitrate to nitrogen gas (N_2), which is then off-gased to the environment. Denitrifying bacteria are heterotrophs and need a source of organic carbon to survive. Frequently methanol and ethanol are used as cheap carbon sources. A denitrifying biofilter can use the same plastic media as an MBBR. The challenge of designing an efficient denitrification system is finding a way to keep the water moving over the biofilter media without increasing the dissolved oxygen concentrations. This can be accomplished by putting a lid on the biofilter to reduce gas exchange with the air, and by using a propeller powered by a motor to mix up the media. Another common denitrification biofilter consists of a tank filled with a static bed of wood chips. The water slowly flows between the wood chips while the oxygen concentration is lowered by the respiration of other

Chapter 6: Additional Equipment Considerations

aerobic bacteria, and the wood chips provide a source of carbon for the denitrifying bacteria. The final method is to send the water to a constructed wetland where most of the nitrate is removed from the water by natural wetland grasses. This method also helps with phosphorous removal since the three most important nutrients to plant growth are nitrogen, phosphorous, and potassium. Phosphorous can also be removed through the use of iron oxide beds and other chemical processes, but generally these methods are overkill for aquaculture wastewater treatment. In general, none of these methods are cost prohibitive to implement or operate. All of the above described methods can also be used on saltwater systems, although in the case of the constructed wetland the plants would need be salt tolerant.

A final alternative for wastewater treatment is the use of a hydroponic system. Similar to a constructed wetland, the plants in the hydroponic system will remove the nitrate and phosphorous from the water. The advantage of this method is that at the end you have a harvestable crop that can be sold. The disadvantage is that the construction of a hydroponic requires investment in greenhouses and additional equipment.

Constructed wetlands can also help with the removal of TSS and TDS. The aquaculture solids are largely composed of carbon that can be consumed by heterotrophic bacteria that live in the soils of constructed wetlands. Solids can also be treated so that they can be turned into a useful and valuable byproduct. This treatment only involves a single step known as dewatering or sludge thickening. When aquaculture solids

Chapter 6: Additional Equipment Considerations

leave the farm they are typically only 0.001-0.005% solids by weight, the rest being water. Solids need to be dewatered to remove the majority of the water weight if you wish to transport them or use them as fertilizer. Dewatering can be accomplished through a variety of methods including belt thickeners, drum thickeners, gravity thickeners, centrifuges, or filter press. All methods do a good job at getting the solids to a concentration of approximately 8% solids by weight. At this point the solids are concentrated but can still be pumped with sludge pumps. Thickened solids can be shipped in tanker trucks to agricultural areas where the solids can serve as an organic fertilizer.

Constructed wetlands are a low cost method for wastewater treatment if space allows. Explore other options if only a smaller footprint is available. Regardless of how the wastewater is treated all parameters need to be able to regularly meet the NDPES target levels. Regulations often require regular self monitoring and reporting of wastewater discharge parameters. This typically involves weekly testing and sometimes the installation of datalogging electronic sensors in the discharge pipe. If target levels cannot be met regularly the wastewater treatment system will need to be modified. It is better to design the system correctly the first time by making the effort to identify potential pollutants and their estimated concentrations before construction begins.

Water Supply Treatment

One of the goals of RAS is to reduce the water needs of farming fish, but water has to initially go into the system and some new water is still needed on a daily basis for replacing

Chapter 6: Additional Equipment Considerations

dirty and evaporated water. Ideally your chosen water supply will not require any treatment before it enters the farm, however this is often not the case. It is important that the water supply is as clean as reasonably possible before it enters the farm. This means off-gasing any excess gases such as nitrogen or carbon dioxide, killing any pathogens, filtering out solids, adjusting pH, alkalinity and temperature, and removing nitrates, phosphates, heavy metals, and excessive hardness. All water that enters the farm needs to be treated, this includes water used for washing, transporting fish, or replacing system water.

In order to determine if any treatment is needed, a comprehensive water test should be performed on all water sources, this includes testing wells of different depths, city supplied water sources, or open water reservoirs. Consult with a local hydrologist or environmental engineer to make sure the water source is of sufficient volume and quality all year round. Water testing will indicate what parameters need to be adjusted before the water enters the farm.

The goal of water supply treatment is to avoid burdening the RAS water treatment equipment and to adjust any parameters that the main RAS water treatment system is not capable of handling. If you have selected a poor quality water supply the treatment system to get the water up to standard can end up using 20% of the entire farm budget; all the more reason to choose a high-quality, consistent water source (See Appendix G). Avoid water sources with excessive minerals or water supplies that are contaminated by agriculture run offs. Drinking

Chapter 6: Additional Equipment Considerations

water standards and treatment methods are similar to what is needed for aquaculture supply water. Drinking water resources can help you determine what methods may be needed to treat your water supply.

Common parameters that can easily be adjusted are temperature and pH. Temperature is adjusted with a heater or chiller, and pH can be adjusted with carbon dioxide strippers or the addition of carbonates. Solids can be removed with any kind of solids filter that is appropriate for the flow rate. All incoming water needs to be sterilized by UV or ozone filters in order to maintain biosecurity. The salinity of the water can be increased with a salt mixer, but it is prohibitively expensive to remove salinity from the water. Similarly, removing heavy metals, phosphates, nitrates, and hardness require sophisticated, expensive equipment that should be avoided. Most of these substances can be removed through the use of activated carbon or ion exchange filters. However, all of these filters require the media to be regularly refreshed or replaced. A city water utility has the money to do this, but the average RAS operator does not have the cash flow to justify treating all of their incoming water this way.

The water supply treatment system can be operated in a couple of different ways depending on how the water is distributed to the farm. It is best to have the system continuously treat a consistent water flow rate and store any extra treated water in a tank until it is needed on the farm. This makes sizing the equipment easier, since it does not need to meet surge demands, such as when a tank needs to be quickly

Chapter 6: Additional Equipment Considerations

refilled; the stored water that was already treated can instead handle the surge.

Water supply treatment ensures diseases and harmful substances are not entering the farm. It also ensures that excessive nitrates or carbon dioxide are not being added to the system, and the pH and temperature are in line with the system water. Budget appropriately for a water supply treatment system to guarantee your farm always has high-quality water on hand.

Oxygen Generator

An oxygen generator is a must have for almost any sized RAS. Oxygen generators are typically rated by the LPM of oxygen that they produce and they range in size from a suitcase sized 10 LPM units to shipping container sized 6000 LPM units. Smaller units supply the oxygen at lower pressures of approximately 1 bar with larger units able to deliver pressures of 6 bars or more. An oxygen generator needs to be able to supply enough pressure to supply all of the oxygenators in the farm. The pressure can be calculated by calculating the head loss from the oxygen piping and adding that to the internal pressure of the oxygenators. The pressure needs to be at least 0.5 bars above the calculated pressure in order to be able to adequately supply oxygen. The flow rate of the oxygen generator should be high enough to meet the maximum oxygen demands of the farm during peak biomasses. The gas supplied by oxygen generators is typically between 90 and 97% oxygen in purity, and the exact concentration should be

Chapter 6: Additional Equipment Considerations

considered in your sizing calculations. Similarly the dissolution efficiency of the oxygenators needs to be considered to determine the total flow rate needed.

The name oxygen generator is a misnomer since they do not actually generate oxygen from anything; instead they merely concentrate the oxygen that is already present in air. This is accomplished by removing the nitrogen from the air through the use of a molecular sieve under high pressure. Rapid pressure swings allow the nitrogen to be preferentially adsorbed by zeolite in a separate vessel. The vessel full of nitrogen can then be isolated with a valve from the oxygen concentrated air, which is sent to an oxygen holding tank. The nitrogen is vented and the process is repeated with a new batch of air. There are two commons methods of removing nitrogen from air, vacuum swing adsorption (VSA) and pressure swing adsorption (PSA). VSA is the newer and superior technology that is more energy efficient than most PSA generators. Both methods operate under the same principle though and smaller units typically use PSA to concentrate the oxygen. An oxygen generator should have an efficiency of approximately 1 Kw-h per cubic meter of oxygen produced. This comes out to $0.05- $0.10 per kilogram of oxygen for common US power rates.

Most oxygen generators are available in variety of voltages and the larger ones operate on 230V power. The generator needs to be located out of the elements but also in a location where it is well ventilated since it is regularly purging high concentrations of nitrogen back into the atmosphere. All generators require regular maintenance such as replacement of air filters and occasional replacement of the zeolite molecular sieve. A number of manufacturers market oxygen generators to the aquaculture industry and all are good at supporting their products to ensure that they have a long useful life.

Ozone Generator

Ozone generators come in a variety of sizes and are typically specified based on the grams of ozone per hour that they produce. A small generator would be 4g/hr and a large unit would be 50g/hr or greater. At larger sizes a manufacturer will be able to produce a custom skid that produces the exact amount of ozone that you require. They do this by adding identical smaller ozone generators to a single skid. Ozone is most commonly generated through a method known as corona discharge where oxygen is subjected to high voltages that break the molecules apart. Using this method 3 to 10 percent of available oxygen is converted into ozone. As a result, using pure oxygen as the feed gas into the ozone generator will yield a lot more ozone than using air alone. Oxygen should always be used with an ozone generator if it is available on site. Regardless, the feed gas will need to be slightly pressurized via a compressor. If air is the feed gas an air dryer may be needed first to reduce the humidity.

Larger generators are more efficient, with more grams of ozone generated per watt of energy used. Ozone generation creates a lot of heat waste and fans are needed to cool the ozone blocks where the corona discharge takes place. In larger units

Chapter 6: Additional Equipment Considerations

water cooling may be required. All units include a gas flow meter and they should be integrated with the correct ORP probes and controls to turn the unit off if redox potential becomes too high in the system. Ozone generator use is becoming more common in aquaculture and manufacturers are recognizing that it is a growing market that requires custom equipment. A rough estimate shows that 250 MT/year of production would require a generator that can produce 500 g/hr, which would require around 5 kilowatts of energy with a capital cost of approximately $40,000. This is not cheap from an operational or capital investment point of view, but its cost will be offset by increased production due to improved water quality.

Emergency Oxygen

Emergency oxygen is needed for all RAS as an insurance against catastrophic failures and it has the dual function of being able to support higher stocking densities. Essentially, emergency oxygen serves as a double back up in case of power failure. The first line of defense is the generator, but the second line of defense is needed in case the generator is delayed in turning on or fails for some other reason. In most RAS there is only enough oxygen in the system water at any one time to support the fish for 15 minutes or less if water circulation were to cease. Oxygen depletion will kill all of the fish on hand in minutes if not properly handled, and since the value of fish in the system is typically equal to the value of the entire RAS system it makes sense to invest in an emergency oxygen system to keep them alive in case of disaster. The

Chapter 6: Additional Equipment Considerations

emergency oxygen system will buy hours of time to fix the problems with the power supply, generator, or any other issues that may be occurring. The system also gives you the ability to cut off the main water flow for short periods of time in case emergency maintenance needs to occur on pumps, solids filters or other water treatment processes. It also can be used to supplement the main oxygenation system when oxygen demands are extremely high such as just after feeding in the weeks before final harvest when the stocking density is the highest.

A typical emergency oxygen system consists of airstones and a liquid oxygen tank. The pressurized liquid oxygen can operate without power and the airstones do not have any moving parts that can fail. The airstones should ideally be in the tanks at all times so that they can be immediately activated if water flow ceases. The airstones can also hang on the side of the tank and be tossed in manually within minutes of failure. And a final option is to have the airstones suspended above the tanks and held up by electromagnets that will release in the event of the loss the power and drop the airstones into the tank. All emergency oxygen systems should have a degree of automation. They can automatically trigger when oxygen levels become too low, however this requires electronic sensors that may not work when power fails. The alternative is to have the emergency oxygen connected to the LOX tank with a solenoid valve that automatically opens in the case of power failure. This ensures that in the event of any power failure the emergency oxygen will immediately kick in.

Chapter 6: Additional Equipment Considerations

The emergency oxygen system should be sized for the highest stocking densities and maximum oxygen consumptions that can be expected in the system. In deep tanks of three meters or more in depth 40% dissolution efficiency can be assumed if fine bubble oxygen diffusers are used. Shallower tanks will have dissolution efficiencies of 30% or less. Airstone manufactures will provide flow rates, oxygenation capacity, and dissolution efficiency in relation to depth for their products so that the correct number of diffusers can be chosen. Similarly, the liquid oxygen tank needs to be large enough to provide at least a few hours of oxygen to the entire farm at maximum flow rate. Of course a bigger LOX tank is always better, but a large tank may become cost prohibitive and typically has be placed outside along the side of the building where it can be easily refilled by a large truck. All components of the emergency oxygen system need to be oxygen safe and corrosion resistant. All safety measures for working around oxygen should be observed.

The emergency oxygen system should be regularly exercised and tested to make sure all components are working. A drawback of emergency oxygen systems is that airstones that continuously sit in the tank can easily become biofouled and will collect solids around them; this is the advantage of keeping stones out of the tank until they are needed. Stones left in the tank should be cleaned on a regular basis. Despite the extra maintenance needs, an emergency oxygen system is a must have for every commercial RAS. Those who do not bother to initially install one come to regret it after their first major fish die off due to a simple pump failure or other system failure.

Chapter 6: Additional Equipment Considerations

Fish Handling

It is good practice to plan for how fish will be moved between systems during the farm design process. Many times it becomes an afterthought and there are then challenges because of the distances between systems, or obstacles in the way. If a farm has been designed properly, fish will not have to travel great distances between systems and there will be a clear path for them. Similarly, it should be easy to deliver fingerlings to the nursery for the first time and to remove harvested fish that will be sent on delivery trucks. Fish handling is also done periodically in order to grade fish by size. All of these needs require an efficient, well thought out, and comprehensive fish handling system and methodology.

There are a couple of different methods for handling fish depending on how much manual labor you want to do. The pure manual labor method involves using nets and bins exclusively. Seine nets can be used to crowd the fish towards one side of tank where dip nets can then be used to easily scoop fish from the tank and place them into another tank or a wheeled bin. When a lot of fish need to be moved this becomes literally backbreaking work. Dip netting is good for doing spot-checking or for moving lightweight fish around the nursery but it is an unrealistic method to rely on for moving large amounts of heavy fish around the farm.

Fish pumps are the solution to reducing manual labor for all handling needs. Fish pumps literally suck up and move fish along a large flexible pipe before spitting them out into a fish grader, processing table, or tank. A fish pump looks like a giant

Chapter 6: Additional Equipment Considerations

centrifugal pump, the impeller is designed to move water and fish along without causing any harm. Fish pumps typically have a maximum NPSH of around 3 meters and a total head of 10 meters, more than enough to move fish out of deep tanks in a level RAS. Fish pumps are usually specified for the size of fish they will be handling, and in large farms it may be appropriate to have a couple sizes of fish pumps, one for juvenile fish, and another for harvest ready fish. Large fish pumps can be difficult to move around and get in position beside a tank. The pump itself is often on wheels but the large pipes needed on the inlet and outlets are heavy and inflexible.

To reduce labor, each fish handling event should be planned to coordinate with any other fish handling that may need to occur. This reduces the need to drag tubes up and down the farm continuously. A fish pump may still require the use of a seine net to crowd fish into a section of the tank. Alternatively the tank can be drained down to a lower level and the pump inlet inserted into the remaining water to begin sucking up fish. It is also possible to design an entire network of underground pipes exclusively for the use of moving fish around the farm with a fish pump. This allows fish to be easily shuttled from one tank to another through the bottom drain or a side hatch. It is important to keep in mind that a significant amount of water is moved along with the fish. Finding an effective way to not waste the water that is moved with the fish can be a challenge. This water can be discharged into the tank where the fish are being moved or it may need to be pumped back to where it came from at an equal rate with another pump. In the case of

Chapter 6: Additional Equipment Considerations

fish processing or grading, the water will need to be drained away and then returned to the tank or system sump.

Most fish species require size grading at least a few times during their life cycle. Grading separates the faster growing, larger fish from the slower growing, smaller fish. This is done since larger fish will often outcompete the small fish for a majority of the food, further widening the size gap between the two and leading to an uneven crop of fish. Grading occurs by moving all of the fish through a grader. A grader consists of v-shaped channels that smaller fish fall through and larger fish do not. The channel may gradually widen along the length of the grader allowing fish to be separated into three or more size ranges. The channel is continuously bathed in water and slightly sloped to keep fish moving along its length. The grader is often coupled with fish counters that register each fish that passes through them. This way the operator knows how many fish were separated into each size class. As fish leave the grader they are sent along an outlet pipe into the appropriate tank. Fish grading causes the fish a great deal of stress and as a result it has to be limited to the minimal amount of times that is needed to keep the farm running efficiently. The entire grading process is typically powered by the use of a fish pump that delivers the fish from the tank to the top of the grader. Similar to the fish pump, the correct grader will depend on the size of the fish you are separating. Each grader works for a range of sizes but a couple may be needed to cover the fish's life cycle from the nursery to final harvest.

Chapter 6: Additional Equipment Considerations

The perfect fish handling formula does not currently exist. Designers and operators use a variety of methods to get the fish where they need to go. Some have elaborate piping systems that allow fish to be quickly moved anywhere in the farm, and others rely on using a dozen people and lots of nets. At the very least, the system should be laid out so that a fish pump can be maneuvered around all of the tanks. And tanks should ideally have built in inlets and outlet ports that fish can be moved into and out of via a fish pump. Moving fish does not have to be a hassle if the proper planning steps are taken during the design process.

Fish Processing & Transportation

Fish processing can take many shapes and forms depending on how much processing actually takes place on. Currently many RAS sell live fish to specialty, niche markets. In this case no actual processing takes place instead fish are moved out of the purge system into holding tanks on the back of a fish transporting truck. These transportation tanks are completely closed and filled with clean water. Pure oxygen is continuously bubbled into the tanks and sometimes zeolite bags are added to the tank to remove ammonia for longer hauls. Ideally the fish are destined for nearby markets that can be reached by a two-hour drive; healthy fish should have no trouble surviving a trip of this distance. Often the farmer is paid based on the weight of live fish that physically make it to the market, any dead fish are wasted profits. The key to live fish transport is to lower the amount of stress placed on the fish at each step. This may also mean lowering the water temperature to reduce the fish's

Chapter 6: Additional Equipment Considerations

metabolism or using an anesthetic that reduces their activity and respiration. Live fish hauling requires some trial and error to find what methods results in the fewest fish mortalities.

If the fish are not destined for the live market, some degree of processing will be required. First the fish should be killed humanely, common techniques include stunning, pithing, and gill cutting. Stunning involves striking the fish's head with a blunt instrument; this can be done manually or automatically. Pithing involves using a sharp spike to strike the fish's brain, this is most commonly done with a specially designed equipment. And gill cutting involves cutting the gills to bleed the fish. Rapid killing is not only humane but it avoids the build up of lactic acid in the fish's muscles that can result in undesirable flavors. The use of automated stunners or bleeders should be used at all farms not selling live fish.

After being killed the fish should be immediately placed into an ice slurry. An ice slurry can rapidly lower the temperature of the fishes meat and hold it near freezing. Temperature control during each of the processing steps is crucial to maintaining high flesh quality. Once a fish leaves the farm the meat will be continuously held at a temperature near freezing, until it is purchased or delivered to the final customer. This journey is commonly referred to as the cold chain, and the farm is the first link in this chain.

After a fish has been killed the next level of fish processing would be head on gutted (HOG). This is commonly done with salmonids soon after they are killed. It involves slitting the

Chapter 6: Additional Equipment Considerations

abdomen and removing the fish's digestive and sexual organs. At this point the eviscerated fish can be packaged and shipped fresh or frozen to supermarkets. The HOG fish can also be further processed into a number of products such as pan fish, where the head, and fins are all removed, or even further processed into boneless fillets. Additional options include pre-seasoned frozen products, or steak cuts. The final product you choose will depend on your marketing strategy.

Fish processing facilities use a lot of specialty equipment along with a large number of low-wage employees to cost-effectively process the fish. The facilities must also comply with federal food safety rules and regulations and are subject to regular inspections. Due to the economies of scale for fish processing it makes sense for the fish processing to be done by a third party rather than the farmer.

Fish processing and transportation is important for RAS operators since most are attempting to market a premium, fresh product. Identify what degree of processing or preparation needs to take place on site and invest in the appropriate equipment to make the killing and movement of fish an efficient process.

Purging

Purging is the process of holding fish in clean water for 4-7 days so that they can eliminate off-flavor compounds from their meat (also see Purging section in Chapter 7). These off-flavor causing substances are present in many systems that operate

Chapter 6: Additional Equipment Considerations

with a biofilter. Purging systems are currently needed for all fish species that are grown in RAS, although research is currently underway to eliminate the need for purging. The system itself is typically a large tank that is operated as either a flow through or partial reuse system. This is done since no biofiltration is allowed and ammonia levels must still be maintained below toxic levels, fortunately the fish are never fed while in the purge system, reducing the ammonia loads. Typically the new water that enters the farm makes a first pass through the purge system before being delivered to the main system sumps. In addition to new water being continuously added to the tank there is often time also supplemental aeration to serve the dual process of oxygenating and carbon dioxide stripping.

The purge tanks need to be large enough to hold at least a week's worth of fish deliveries at stocking densities of 50kg/m^3 or less. Multiple tanks may be needed if you have a daily or every other day harvesting schedule, and all of the tanks should be located near the processing and transportation areas of the farm. Aeration can be accomplished with either airstones or a stripping tower, both can be sized for a reduced oxygen consumption load since no feeding is taking place. The purge tanks should have an emergency oxygen system like the rest of the tanks. Monitor the purge system the same as any other system to ensure the water quality is adequate.

Monitoring and Control System

The monitoring and control system is made up of sensors, alarms, automated valves, and electrical switches. It can be

Chapter 6: Additional Equipment Considerations

used to operate emergency oxygen systems, water heater/chillers, pump VFDs, drum filters, and other water treatment components. The goal of the system is to essentially take the place of a hyper-vigilant RAS operator who sees everything at all times and is capable of making quick basic decisions to mitigate problems or warn others of possible issues.

The control system can be built with various levels of complexity in terms of automation. At its most basic a control system consists of various sensors that send signals to programmed alarms and controllers (Figure 15). A simple example would be a thermometer located in the fish tank indicating that the temperature is above the chosen operating range and a signal is sent so that the heater shuts off until the temperature drops back into the ideal range. This system of sensors, alarms, and controllers can be centralized in a single control panel or it can be decentralized with each piece of equipment operating on its own control system, such as the heater example. A centralized control system makes management easier, but it requires a custom built system.

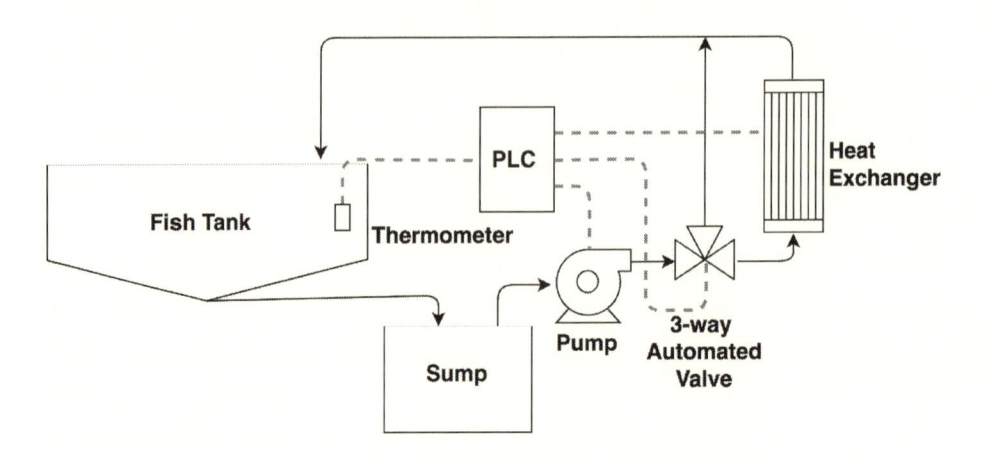

Chapter 6: Additional Equipment Considerations

Figure 15. An example of a heat exchanger side loop with a bypass. Black arrows indicate water flow and the dashed gray line shows electrical connections to the PLC. The thermometer reading determines how the 3-way automated valve is positioned; it either sends water to the heat exchanger or bypasses it.

A control system is often times the technically most complex piece of equipment on any RAS operation and it requires specialist knowledge to design, install, and troubleshoot. The components of a control system are the sensors, logic controllers/computers, alarms, and electrical switches. A piping and instrumentation diagram (P&ID) is used to layout and show how all of control system components connect. Industrial monitoring and control systems are used in most process engineering industries such as wastewater treatment, food processing, and chemical processing. Many of the same components used in those industries can be used in a RAS control system.

Electronic sensors are the frontline of the control system. Common sensors include temperature, pH, salinity, dissolved

Chapter 6: Additional Equipment Considerations

oxygen, water level, redox potential, and flow rate. All of these sensors use different methods to turn a physical parameter into an electric voltage. The magnitude of that voltage determines the parameter's value, a basic circuit board can transform that voltage into the parameters desired units and display it to the operator. The relationship between the voltage and the parameter's value is calculated by a previously determined empirical equation. In some cases, such as pH, salinity, and redox potential, the voltage can be read out directly and interpreted by a trained operator. In other cases, such as dissolved oxygen, temperature and salinity must also be known in order to accurately calculate the DO in mg/L.

All sensors have varying levels of precision and accuracy. Precision is the resolution at which the sensor can detect changes and accuracy is how closely the measured value reflects the actual value. Pay careful attention to the precision and accuracy when selecting your sensors, typically sensors with increased precision and accuracy cost a lot more, but the added utility of having them is negligible. Similarly, different sensors require different amounts of care, maintenance, and calibration. Electrochemical sensors often have to be calibrated or recharged somewhat frequently so that they remain accurate, while optical sensors just need to be wiped down periodically. Water level sensors are particularly robust and unlike other sensors which give a range of values they instead are typically binary, the water level is either too high/low or there is nothing to report. Sensors are just one component of a control system, but they are the piece that varies the most in price and quality. Take the time to educate yourself on the

Chapter 6: Additional Equipment Considerations

advantages and drawbacks of each sensor and understand how it will be applied in your farm.

The brain of the control system is a programmable logic controller (PLC). A PLC is an industrial computer that is rugged, reliable, inexpensive, and easy to program. The PLC talks to all of the control system components and it is programmed to make decisions based on the feedback that it receives from the sensors. The decisions are pre-programmed in the PLC and based on what it hears from the sensors it can trigger alarms, valves, or pumps. Multiple PLCs can be mounted in a weatherproof electrical control panel cabinet. In the panel they are connected to a power supply, relays and controllers. The PLC can also connect to a desktop computer or standalone hard drive to function as a datalogger, regularly recording and saving all the sensor's values and statuses of equipment. The PLC is a crucial component of a comprehensive supervisory control and data acquisition system (SCADA). A SCADA system adds another level of control, communication, and complexity. Typically a SCADA system will include a graphical user interface that makes it easy to visualize and navigate all of the available data coming from the system's sensors. Most new RAS farms use SCADA software, which is available from a number of providers, to manage the entire monitoring and control system.

Alarms are used in even the most basic control systems; they are used to signal important or urgent water quality or equipment issues. These include low dissolved oxygen, electrical outage, low water flow, high/low water level, and flooding. The alarm itself can be an audible siren and flashing

Chapter 6: Additional Equipment Considerations

light, or more commonly a mobile phone is called and a code is read out that explains the reason for the alarm. The phone calling system makes it so that the farm operator is alerted even if they are offsite or away from the equipment. All alarms are preprogrammed to trigger when a sensor reads a value outside of its range. An example would be having an alarm trigger whenever an in tank oxygen sensor reads a concentration of 5.0 mg/L or less. Unfortunately, alarms will also often trigger when sensors have false readings or other errors.

In addition to triggering alarms the other goal of a control system is to actually control things on the farm such as pumps, heaters, emergency oxygen systems, and most other equipment. Pumps can be controlled with VFDs that will increase or decrease the water flow rate by managing the electrical power. This can be helpful when the head loss varies but you still require a constant flow rate. For example, as the solids build up in a bead filter the head loss will increase and the flow rate will begin to decrease. Instead the flow rate can be held constant by increasing power to the pump through a VFD that is receiving feedback from a water flow sensor that is communicating with a PLC.

Automated valves are another method to control water flow. Based on what water flow and pressure sensors are measuring, valves can be shut or opened to regulate where and when water flows to various pieces of equipment. Automated valves can help with the regulation of the heating and cooling system or with the backwashing of a bead filter. The control system can

Chapter 6: Additional Equipment Considerations

be used to automatically turn on the emergency oxygen system based on what the dissolved oxygen sensors are reading. Drum filters are always operated with a control system that will start the drum turning and turn on the pressure pump as soon as water behind the drum filter begins to back up and a water level sensor is triggered. These are just a few ways that the control system can be used to operate equipment automatically. Automation increases water quality consistency, while reducing the need for manpower, the response time for emergencies, and the farm's energy usage.

The control system is designed to create a feedback loop where the sensors gather data and feed that data back to the PLC, which then makes decisions and takes the appropriate actions by triggering alarms or turning on/off equipment. A nearly countless number of these loops can be operating simultaneously and the complexity of the decision-making process is only limited by the skill of the programmer. Technological advancements such as inexpensive microcomputers are making control systems increasingly cost-effective and they are eliminating the need for manual monitoring and operation of some RAS equipment. Automation and complex control systems will play a key role in all future RAS operations.

Generator

A diesel back-up generator is a necessity for all RAS given the high probability of the power failing at some point. A generator is cheap insurance against the possibility of loosing all of the

Chapter 6: Additional Equipment Considerations

farm's fish due to a natural disaster or other hiccup that causes power interruption. Diesel generators serve as the back up power systems for residential, industrial and commercial buildings. As a result they are available from a number of manufacturers and distributors in a wide range of sizes. The generator needs to provide enough power to keep all of the farm's main treatment systems online; this includes pumps, blowers, oxygen generators, drum filters, carbon dioxide strippers, and lighting. Heating/cooling systems can be operated at a reduced load for a short period of time and other equipment such as fish pumps do not need to be operated during a power outage. The generator is set up to start automatically as soon as the utility power fails with full power generation typically reached around 15 seconds after power failure. It is possible that newly available large battery storage systems could be used to bridge the gap between utility failure and generator start up, insuring non-stop power supply.

Generators are typically tested once a week by starting the generator up and letting it run for a half hour to make sure it is in good operating condition. The diesel storage tank needs to hold enough fuel to keep the generator operating for a few days in case of catastrophic power failure and the tank will need to be topped off a few times year. The generator should be located alongside the RAS building where it can be serviced and the diesel storage tank can be refilled. Generators are a significant cost, but they have a long life and provide all farms with a reliable method of maintaining continuous electrical power for the water treatment system.

Chapter 6: Additional Equipment Considerations

Building

A good RAS building makes everything easier. It provides temperature stability, creates a biosecurity barrier, allows fish and feed to come and go from the farm as needed, and provides the space and services for personal. The ideal RAS building is inexpensive, corrosion and mold resistant, well insulated, and easy to wash down. This rules out the use of wood, unsealed insulation, and untreated steel. Aluminum, galvanized/coated steel, and plastic are the materials of choice for walls, cladding and support beams. There are a number of manufacturers who make aluminum clad pre-fabricated buildings at a reasonable price that work well for covering a RAS. Concrete is another great material choice and it is the best choice for the entire floor and foundation. Concrete does not corrode and is easy to wash. Dirt or gravel floors are difficult or impossible to sanitize, and plastic or wood floors break down over time with continual exposure to water.

Insulation is a worthwhile investment for most RAS operations; the need ultimately depends on the local climate. Insulated walls and roofs significantly reduce the heating and cooling costs of the farm. However, given the high humidity inside the farm, solid foam insulation or similar materials is highly recommended. A plastic membrane can also be used to essentially seal the air (and moisture) inside of the building so that the building materials and insulation are not exposed to the humidity. This also reduces drafts and allows for total control of the ventilation rate.

Chapter 6: Additional Equipment Considerations

The building needs to be positioned and designed so that people, feed, and fish can easily come and go from the building, while still maintaining biosecurity. Farm personnel can enter and leave via one or two standard sized doors, and fish and feed can be moved into and out of the farm through roll-up service doors. Ideally trucks can back up to the service door so that materials can easily be loaded and unloaded. Walls inside of the building can be used to separate systems or storage areas. In addition to housing the RAS and storage space the building should also have space for offices, a break area, and washrooms. It is also common for these facilities to instead be housed in a nearby adjacent building.

It is common for a RAS farm to be housed inside of repurposed existing buildings such as old barns or warehouse. Given the specific needs of a RAS and the large amount of underground piping it is strongly recommended that a new building be used for any new farm. This allows for a lot of the underground work to be installed before the building goes up and the best building materials can be used so the building will out last the farm equipment. Greenhouses are also commonly used to house both small and large RASs. They are often made of corrosion resistant and easy to clean materials such as aluminum, plastic, and glass. And their biggest advantage is that they are inexpensive. However, a greenhouse may be great when a lot of light is needed to grow plants, but for RAS the extra sunlight leads to unnecessary algae growth. Also temperature control is more difficult in a greenhouse since they are uninsulated and cool down significantly during the night. Greenhouses should only be used in specific climates or for

Chapter 6: Additional Equipment Considerations

small inexpensive systems. Commercial RAS should be housed in an insulated, climate controlled building.

The building design needs to meet or exceed local building code regulations. An engineer will have to stamp the design and show that the building is capable of withstanding any possible snow loads, wind loads, or earthquakes. The total building cost will depend on the materials used, the total square footage, local labor rates, shipping costs, and local building codes. Local construction firms and engineers can help with determining what building will ultimately work best for your budget.

HVAC

The heating, ventilation, and air conditioning (HVAC) system works in tandem with the water heating/cooling system to help maintain a stable water temperature by controlling the temperature and humidity of the air inside of the farm building. Temperature is easy to manage with the appropriate heaters and air conditioners, which one you use will depend on your local climate and the temperature of the RAS water. In a cold water system the air temperature is typically held a couple of degrees Celsius above the water temperature, and in warm water system the air is held at or just above the water temperature. These air temperatures can sometimes be uncomfortable for unaccustomed workers, but with the right clothes it is not a problem. Keeping the air temperature near the water temperature reduces the energy loads of the water heating and cooling system. Air and water are continuously

Chapter 6: Additional Equipment Considerations

exchanging heat and the water is continuously evaporating. Warmer water temperatures lead to higher rates of evaporation and evaporative heat losses. Slightly warmer air temperatures can partially compensate for these evaporative heat losses.

Humidity management can be more complex. All RAS lose water to evaporation, and those evaporative losses occur in an enclosed space, which leads to high relative humidity, often near 100%. The problem with a high relative humidity is water condenses on a lot of surfaces such as the ceilings, walls, and equipment. Eventually this can lead to mold growth or hasten the rate of corrosion. In order to keep the humidity levels lower the air can be heated, which reduces the relative humidity, or it can be ventilated out of the building and replaced with lower humidity air from the outdoors. Ventilation is needed regardless since carbon dioxide concentrations will increase inside the farm building if no ventilation occurs. The drawback of ventilation is the replacement air is often not at the same temperature as the farm's air temperature. The optimal ventilation rate is to exchange enough air to control humidity and carbon dioxide levels, but not so much that you have large air heating/cooling energy loads. There is ventilation equipment that will work as a heat exchanger between the incoming and outgoing air to reduce the energy needs of heating or cooling the new air. Regardless, some heating/cooling will need to occur to keep the air temperature within the correct range.

Chapter 6: Additional Equipment Considerations

Piping

RAS piping requires a lot of engineering, planning, and construction time. The challenge of building a good system of pipes is limiting the cost of materials while also reducing the pipe friction head losses. Pipe costs increase dramatically as the pipe diameter increases. Large pipe diameters are needed though to limit frictional head losses for the high water flow rates seen in RAS. As water moves through pipes it rubs up against the pipe walls creating friction. The friction increases exponentially with an increase in the water velocity. Each bend in the pipe or valve also creates turbulence and friction. The problem with all of this friction is it requires energy to overcome those frictional losses and that energy comes from the pumps, which run on electricity, which costs money. If pipes are selected and installed correctly they should each have a water velocity of 2 m/s or less and the number of turns, valves, and constrictions would be limited to only what is absolutely necessary.

$$h_L = f \frac{L}{D} \left(\frac{v^2}{2g} \right) \qquad h_L = K \left(\frac{v^2}{2g} \right)$$

The equation on the left calculates the major head losses in meters of water that is due to the frictional losses from the water flowing through the pipe. Where v is equal to water velocity, g is equal to gravitational acceleration, L is equal to the length of the pipe, D is equal to pipe diameter, and f is equal to the frictional constant. The frictional constant depends on a number of factors including the liquid's viscosity, the roughness of the pipe, and the pipe diameter. f can be

Chapter 6: Additional Equipment Considerations

determined by calculating the Reynolds number and using a Moody chart to look up the appropriate value. Many pipe suppliers also provide ready-made tables that can be used in lieu of doing the calculation. These tables give the expected frictional head loss per one hundred feet of pipe for every diameter of pipe at a range of water velocities. Looking at this equation we can see how water velocity and pipe length increases frictional head losses and how a larger pipe diameter will lower frictional losses.

The equation on the right calculates the minor head losses in meters of water, which are the losses due to turns, valves, or other constrictions. K is a unit less constant that has been experimentally determined for a variety of fittings such as ball valves, ninety-degree turns, or sudden pipe diameter contractions. K values can be looked up for each fitting in a table and the total minor head losses can be added together with the major head losses to determine the total frictional pipe losses.

The system piping can be constructed from a variety of metals and plastics. Almost all piping in aquaculture is either polyvinyl chloride (PVC) or high-density polyethylene (HDPE). Neither leaches potentially harmful substances into the water like metals, nor do they have corrosion issues. Both have adequate pressure ratings for use in RAS, PVC being better suited for higher-pressure applications. PVC is also typically more readily available and cheaper for all pipes and fittings with diameters of 8 inches or less. HDPE becomes cost competitive for larger diameter pipes and particularly for use with custom fittings that

Chapter 6: Additional Equipment Considerations

can be made by local fabricators. Custom fittings are often needed for distribution manifolds or pump connections. Typically a combination of HDPE and PVC makes the most sense. HDPE is used for all of the large pipes handling the main treatment flow and PVC is used on side stream applications or smaller systems. The plumbing contractor will know what material should be used for each application.

Valves are available in a number of different designs and at a huge range of prices. Twelve-inch diameter or larger valves often cost a $1000. If you need a lot of valves to control and redirect flow these costs can add up quickly. Same as the pipe, valves are constructed from various metals and plastics. Large plastic valves (>6") are cost prohibitive and coated cast iron is more appropriate for most applications, where as small valves (<6") can be purchased economically in PVC or similar plastic compounds. Common valve types are ball valves, butterfly valves, gate valves, and swing check valves. Ball valves are the best choice for all smaller valves, since they are mass produced and economical. Butterfly valves are appropriate for larger pipe diameters of 8 inches or greater. Butterfly valves are available in either wafer type or lug type depending on how you want to attach the valve to the pipe flanges. And butterfly valves can also be used with a lever or gear operator, gear operator being the best choice for large valves or in space limited applications. Often times the rubber seal material needs to be specified, and the less expensive option of EPDM is suitable for use with all water applications. An advantage of ball and butterfly valves is that they can be used partially open and adjusted for how much flow you may want between zero flow and full flow. In

Chapter 6: Additional Equipment Considerations

contrast, gate valves can only be used if fully opened or fully closed, but they are an inexpensive option for pipe diameters of eight inches or less. Swing check valves cannot be operated manually but instead automatically close when water flow ceases, this prevents water from flowing backwards through the pipe.

All valve types can be controlled and operated electronically with the use of an actuator. Naturally this adds to the cost of the valve but in some cases it may be appropriate to have the valves operated automatically. Only use what valves are needed for the system to operate properly since they create head losses and contribute to piping costs. Despite this, valves are often needed so that equipment or sections of the farm can be isolated for brief periods of time while repairs or other maintenance occurs. They can also be used to redirect flow in the case of an emergency. Take the time to shop around for valves and find materials and prices that make sense for your operation.

Installation of the system's piping is a considerable expense. For starters, a lot of the piping is buried below grade and heavy machinery is needed to do the digging and placement of the piping. Furthermore, plumbers are skilled tradesmen and as a result they command good wages. However, a good plumber will be able to solve tough installation problems and do the job correctly the first time so that once a pipe is buried and set in concrete it never needs to be dug up. Additional costs include pipe flanges and pipe supports. Pipes need to connect to all of the equipment and tanks; this is typically done with bolted pipe

Chapter 6: Additional Equipment Considerations

flanges. The advantage of using bolted flanges instead of gluing the pipe in place is that the equipment can be easily changed out. Pipe supports are needed for all piping that is not below grade due to the pipe's weight when filled with water. Most pipe supports can be custom fabricated out of corrosion resistant metal. Make sure when estimating piping costs that you include all fittings, valves, flanges, supports, and installation labor.

Biofouling is a potential problem for all pipes that can eventually increase head losses by decreasing the diameter of the pipe and increasing pipe roughness. Biofouling can't really be avoided but it can be relieved by regularly cleaning the pipes. This is most efficiently done with a pipe cleaner, also known as a pig, that is really just a plug that can be either pushed or pulled through the pipe to scrape the walls clean of any biofouling. Pipe cleanouts facilitate the use of pigs by creating an easy spot for an operator to insert and extract a pig. Install pipe cleanouts on all pipes that may be prone to biofouling.

An alternative to pipes, for all unpressurized flow, is open channels. Open channels are simply channels constructed of concrete or aluminum that allows water to flow via gravity through them. Open channels only make sense for use with the main treatment flow where eight inch or larger diameter pipe would be used instead. A typical application for an open channel would be to convey the water draining from the tanks to the drum filter. The advantage of open channels is they can cost less than large diameter pipe and convey water with very

Chapter 6: Additional Equipment Considerations

little head loss. The open channel needs to be sloped slightly to move the water down its length, and the walls need to be high enough to handle surge flows and prevent the loss of water from splashing.

Plumbing of some kind is used in almost all industries, but aquaculture is unique regarding the dense network of pipes that must run between tanks and the filtration equipment. Many of the pipes also have uncommonly large diameters, which can make things expensive. Take the time to properly estimate the piping costs at your farm because it can often times be a surprisingly high number. Use corrosion resistant materials that will not leach harmful substances into the water, and use engineering drawings to double check that everything is installed properly the first time.

Feeding System

A properly specified and designed feed storage and delivery system is needed for all fish farms. Delivering the correct amount of feed each day is key to maintaining a high feed conversion ratio (FCR). The FCR is a driving factor in determining farm profitability since feed is the number one expense of any RAS. An efficient feed system delivers the correct quantity of food at the correct time to ensure the fish are continuously fed to satiation without any food going to waste. This is a complex task and as a result all feeding systems should be closely controlled and monitored by experienced personnel. All feed use should be carefully tracked and compared with growth rates to continuously monitor the FCR.

Chapter 6: Additional Equipment Considerations

There are a number of software providers who sell products that can help with the monitoring of feed usage. Feed use and growth rate data are the key drivers for all decision-making processes. Without this data neither the water treatment nor the feeding system can be optimized.

The quantity of feed that must be delivered to the fish each day depends on the size of the farm, the biomass of the fish on hand, and the daily feeding rate. Commonly, fish are fed 1.5-5% of body weight per day. Fish feed comes in the form of pellets that are available in a variety of sizes and dietary formulations. The formulation will depend on the fish species being grown and the pellet size will depend on the size of the fish. It is common to have three sizes of pellets on hand for fish that are at different life stages. Fish feed is also often available in floating or sinking pellets. Floating pellets are easier to observe and track, but for some species sinking pellets may be the only option. Regardless of the feed type, it needs to be delivered to the fish in a way that evenly distributes the feed across the surface of the tank. This reduces competition between the fish and helps to ensure even growth. The feed also needs to be provided multiple times during the day, sometimes 24 hours a day and other times only during the 16 hours of daylight. Feeding can be done in a number of different ways depending on the size of the farm and the degree of automation desired. These methods include hand feeding, hopper feeders, and a pneumatic feed system.

The simplest method of feeding is hand feeding multiple times per day. This simply involves a bucket of food and a scoop that

Chapter 6: Additional Equipment Considerations

you can use to fling feed into each tank. The problem is that with large farms you need to fling a few metric tons of food per day. This method does not scale well for use in the main grow-out systems. Hand feeding may still be appropriate in the nursery system where the feed loads are a few orders of magnitude lower. Even with hand feeding everything should be tracked, if you are using a bucket you can weigh it before hand and then again after feeding.

A belt feeder is a device that also works well for the small amounts of feed used in a nursery. Food is placed in a thick layer on top of a belt that slowly retracts over the course of 12-24 hours dumping food into the tank below it. Another method is to use automated or demand hopper feeders suspended above each tank. This typically consists of a large 200 to 400 liter plastic bin suspended above the tank that is attached to an electrically operated spreading disc that can evenly distribute food across the tank. Automated feeders regularly disperse feed on a timed scheduled, while demand feeders have a trigger that hangs down into the tank that fish can knock against to disperse more feed. These bins need to be refilled regularly and all of the feeders need to be tracked.

A pneumatic feeding system is the gold standard for all large fish farms including net pen operations. The system consists of a large hopper filled with a few tons of food that is connected through a distribution box to a network of flexible plastic tubes that deliver food to each tank. The distribution box is electronically controlled to open and close gate valves for each of the tubes. A blower provides air to push the food pellets

Chapter 6: Additional Equipment Considerations

from the hopper through the distribution box to the tanks. Once the feed reaches the tanks it may be directly sprayed on to the water's surface or it can refill a hopper feeder. The entire system is sized for the weight of feed that needs to be delivered per hour and for the number of tanks the feed must reach.

The advantage of this system is the limited feed handling that needs to occur. The main hopper only needs to be refilled occasionally using one metric ton bags of feed. Also, the electronic controls allow you to dose quantifiable amounts of feed to each tank at regularly scheduled intervals throughout the day. This system makes tracking feed usage easy and limits the errors that can occur with manual data entry from handwritten log sheets. The disadvantage is the price of the system. Custom sized pneumatic feed systems are typically assembled from off-the-shelf components, which limits their cost to some degree. Regardless, the savings in man-hours outweighs the costs. Another consideration in pneumatic systems is the feed pellet durability. As pellets are moved and knocked around they begin to degrade and fines are scraped from the sides. These fines can not be consumed by fish and instead just add unnecessary proteins and fats to the water that then have to be removed through a combination of solids filtration, biofiltration, and fine solids filtration. A well-formulated pellet will not generate too many fines and will be able to survive the pneumatic feed system. In addition, the entire pneumatic system is designed with sweeping curves and smooth sides to limit the creation of fines.

Chapter 6: Additional Equipment Considerations

In addition to distributing the feed to the tanks the feed most also be regularly delivered and stored at the farm. At a large farm, feed is commonly delivered by the truckload, which holds approximately 25 metric tons of feed. The feed is packaged in one metric ton sized plastic sacks that sit on top of pallets. The sacks can be moved by forklift and lifted by straps on top of the bag to raise the bag above a hopper. Feed is also commonly available in 25 or 50 kg plastic sacks that are stacked on a pallet. These smaller sacks make feed handling more labor intensive, but may be needed at smaller farms or for the smaller quantities of feed needed in the nursery. Once on the farm, the feed sacks should be stored in a dry, covered area that is protected from pests. The feed can also be loaded into a silo, this is typically done by emptying the sacks into a hopper that feeds a screw elevator that delivers the feed to the top of the silo. Silos are typically placed outdoors directly adjacent to the main farm building. The advantage of a silo is that a large amount of feed can be safely stored inside of it and it can send feed directly to the pneumatic feeding system. Ideally the farm will have room to store at least a weeks worth of food on site.

Most farms will end up using a combination of all the feed handling methods. Nursery or fingerling systems can be fed with belt feeders or hoppers and the main grow out systems can be serviced by a pneumatic system. A silo system makes receiving and storing feed easy. Throughout the entire process the feed needs to be tracked; electronically controlled systems make this easy. A good feed handling system will contribute directly to the bottom line of the farm by insuring no feed goes

Chapter 6: Additional Equipment Considerations

to waste. Take the time to design the entire process and invest in the proper equipment.

Salt Mixers

Land-based saltwater systems that are isolated from a saltwater source will require the regular addition of freshwater mixed with salt to replace wastewater. Water loss due to evaporation can be simply replaced with freshwater since the salt still remains in the system. The salt used to make saltwater is not usually pure sodium chloride, instead it is a mixture of other salts and minerals such as magnesium, calcium, potassium, carbonates, bromide, and iodide. Full-strength saltwater is 35 parts per thousand (ppt) also written as 35 g/L. This is a lot of salt when mixing large quantities of water. For example, a relatively small replacement volume of 4000 liters per day requires approximately 140 kilograms of salt to be mixed in to the water each day. This mixing can require a lot of backbreaking labor and time. To make this work easier and ensure that salt is mixed in evenly the salt and freshwater should be combined in a dedicated salt mixing container. Typically this is a cylindrical or square vessel that is filled with water and then salt bags are emptied into it. An easy mixing method is to use heavy airflow through airstones on the bottom of the tank to agitate the water. Salt will still accumulate in dead spots on the tank floor, but it can be mixed in manually with a stirring stick. The salinity should be monitored and when it is properly mixed the water can be diverted into the main system. Unfortunately recovering the salt from the wastewater is not economically viable, it is too difficult to separate the salt

Chapter 6: Additional Equipment Considerations

away from the water and all of the other aquaculture waste products that are being flushed down the drain.

Blowers

Centralized blowers are needed in systems that use aeration during either biofiltration or carbon dioxide stripping. They are also used on smaller nursery or purge systems to provide both oxygenation and carbon dioxide stripping. The size of these blowers will depend on the airflow rate required, and unless the farm is small, there should be two units that run in parallel so that during servicing or breakdown at least half capacity is still maintained for the farm. Blowers range in size from 1 Kw for use on small nursery systems to 40 Kw for large grow-out systems, units up to 200 Kw are available for truly huge systems. A standard rotary-lobe blower can handle pressures from 0-3 meters, a multistage design can manage a pressure of 5 meters. For pressures beyond this a compressor would be needed, although compressors also do not function efficiently at pressures of 5-10 meters, as a result it is better to design the farm for the use of blowers by keeping pressures below 5 meters. Blowers should always be set up with pressure relief valves, pressure gauges, and vibration isolators.

Blowers can generate a significant amount of noise and heat. Smaller blowers, with power ratings of 7.5 Kw or less, are often sold uninsulated; these blowers should be located in a room that isolates the noise from the main farm floor. The obnoxiousness of loud, noisy equipment should not be underestimated. Earplugs will be needed if the blowers are

Chapter 6: Additional Equipment Considerations

kept in areas that workers frequent. Larger, more expensive blowers can be purchased with a variety of accessories such as sound insulated cabinets and touch screen controls. The air leaving a blower is quiet warm, often around 40 °C. In coldwater systems where the excess heat is not needed it can be partially dissipated through a radiator like device. Periodic maintenance of blowers includes oiling moving parts and cleaning or replacing air filters.

Lighting

All aquaculture systems require some type of lighting system. Light helps fish find food, allows for photoperiod control, and lets workers see what they are doing. Generally farms are located within windowless buildings since sunlight can lead to undesirable algae growth and it makes photoperiod control impossible. Ceiling mounted lights that provide even lighting throughout the building are ideal. Any new farm should invest in LED light technology. The higher priced bulbs do not need to be replaced as often and require significantly less energy per unit of light. Light fixtures should be built for high moisture environments out of materials that do not corrode. The electrical wiring for lights should be similarly moisture resistant.

When photoperiod manipulation is required, the lights will need to be coupled with a controller. Suddenly turning lights on in the morning can frighten and stress fish. A dimmer allows lights to be turned up slowly over the course of thirty minutes, reducing the chances of the fish becoming stressed. Ongoing research is looking at what minimum light levels are needed to

Chapter 6: Additional Equipment Considerations

have the proper photoperiod manipulation effects or if certain wavelengths of light are more crucial than others. Sunlight is 10 to 100 times brighter then even the most well lit buildings and the wavelength spectrum of sunlight cannot be exactly replicated by any artificial light sources. Nonetheless LEDs provide the best alternative for maintaining high lights levels with a comparable wavelength spectrum. It is possible that in the future, an ideal wavelength spectrum will be identified for fish culture and manufacturers will be able to create LEDs that only give off this wavelength. In the meantime a good white light is suitable. Some fish can be grown with lights on 24 hours a day, but to conserve energy the lights can be dimmed at night when less work is going on at the farm. Central light controls are better then spread out switches. Emergency lighting should be connected to the backup generator in case of power loss.

Chapter 7: Operational Considerations

Cleaning

A regular part of system upkeep is the cleaning of all equipment, tanks, and pipes. Over time, nearly every part of the farm that is exposed to water will become covered in a bacteria mat, this is commonly referred to as biofouling. Biofouling is not only aesthetically displeasing but it can make solids removal more difficult and in extreme cases restrict flow in pipes or channels. Biofouling can often times be cleaned from smooth surfaces with a simple brush or broom. The bacterial mats will then become suspended in the water and either be eaten by the fish or filtered out through the solids filter. A deeper cleaning can be done if a system is taken offline and allowed to dry out. The system can be scrubbed cleaned and sterilized with a bleach solution. It should be noted that bleach should be used extremely cautiously around a fish farm. The chlorine in bleach is extremely toxic to fish and bleach should only be used on systems and equipment that will not come in contact with fish until they have been well rinsed. Biofouling can also occur inside of pipes; high enough water velocities inside the pipe will reduce biofouling, but occasionally pipes may need to be cleaned to unclog them of bacteria mats. This can be done by running a pipe cleaner, sometimes called a pig, through the pipes. Some farms regularly clean their pipes to make sure they are free of obstructions and to reduce pumping head. The rate of

Chapter 7: Operational Considerations

biofouling on a farm is largely a function of temperature, with bacteria growing much more rapidly in warmwater systems. There is little that can be done to prevent biofouling. Dead spots in sumps and other vessels can easily fill with solids and should be avoided when designing and constructing all sumps, biofilters, and any other tanks. Certain pieces of equipment may require additional cleaning; these include filter screens, airstones, and sensors. Regular checking and cleaning will prevent potential failures or a reduction in efficiency.

Dry parts of the farm will also need to be regularly cleaned. Floors often get covered in errant fish food, dust, dirt, and other debris. Fish food in particular can become gross when it gets wet due to water splashing on to the floor or by being exposed to a humid environment. Food waste should be avoided and quickly cleaned up if a spill occurs. Brooms, hoses, and mops, are all good tools for keeping floors and other farm areas clean and organized. Keeping the farm clean lessens the chances of having to deal with any pests or bugs. Flies can become nuisances in warmwater farms that are not kept clean, and loose food pellets can attract mice and rats.

Removing dead fish is also a regular part of system cleaning. Fish will likely die every day in a large farm and their bodies will need to removed from the top of the tank, the bottom drains, or the floor since fish will jump out occasionally. All dead fish (often called morts) should be recorded as a part of daily record keeping. Morts left in tanks will quickly decompose and they increase the chances of disease spreading within the system, also they smell bad. As a result they should be

Chapter 7: Operational Considerations

removed promptly and disposed of in a composter or buried on site to reduce smell and discourage scavenging animals.

In general, cleaning should be seen as a part of regular system maintenance and biosecurity. A clean, organized farm reduces wear on equipment and makes all other operations run more smoothly. A farmer can be proud of their clean farm and show it to potential investors, partners, or customers without fear of embarrassment.

Monitoring & Water Testing

Water quality monitoring and testing is a part of every day practices at a fish farm. The electronic sensors of the control and alarm system can do a lot of the work especially if they are connected to a datalogger that regularly charts the ups and downs of temperature, dissolved oxygen, conductivity, and pH. However, as mentioned in the water quality chapter, a number of parameters must be regularly monitored by chemical analysis.

Water sampling typically involves taking a flask of water from each independent system and testing for each of the important parameters. These parameters include TAN, nitrite, nitrate, alkalinity, hardness, and TSS. pH, carbon dioxide, and TDS may also need to be tested for occasionally. There should be a dedicated lab bench somewhere in the farm facilities so these tests can be performed in a comfortable setting. There are a number of different test kits and brands that can be used to test for all of these water parameters. Different methods of

Chapter 7: Operational Considerations

measurement for each parameter have different levels of accuracy and precision. The steps of the water testing kits must be followed carefully in order to ensure precision of the measurements. Also it is common to have the same employee perform the water testing each day or week so that the same methods and techniques are used consistently.

Water testing will need to occur more frequently during farm start up and especially until the biofilters and biomass of the system reach a steady state. Once the farm has been operating without problems for a significant period of time, the water testing can take place less frequently since parameters are less likely to change. If there is ever a rapid deterioration of health observed in the fish or increased mortalities, then water testing should be one the first steps taken towards diagnosing the problem. All data that is collected during water sampling should be recorded and placed in a centralized spreadsheet or database.

Fish Handling

Fish handling can be one of the most labor-intensive jobs on the farm. Limiting the amount of times that fish have to be handled saves the farmer time and effort, and it saves the fish a lot of stress. However, there are certain times that fish must be handled in order to make sure the farm is operating efficiently.

Fish need to be handled for a number of different reasons. First, the fish must initially enter the farm. Since at this point the fish are fairly small the handling is relatively easy and can be

Chapter 7: Operational Considerations

done with tubs or buckets. Once fish are in the nursery and begin growing they are now a part of the farm and will need to be graded periodically to keep similarly sized fish together. Grading is needed for a variety of reasons and the frequency of grading often depends on the fish species. Grading is where most of the handling occurs during the fish's time on the farm. Having the proper fish handling equipment is the easiest way to cut down on the human labor and fish stress. During the grading the fish will most likely be moved from tank to tank. Fish eventually move from the nursery into the grow-out tanks. And from there they may be graded a couple times into other systems and grow-out tanks. Finally, once fish are ready to harvest they will need to be moved into the purge system, and then from there to the processing line or into a truck for shipment to a customer. Fortunately, moving the dead fish after processing is a lot easier than moving live fish since the water is no longer needed and the stress of the fish no longer matters. But many RAS operations get a premium for selling live fish to markets and lowering the stress of the fish is key to increasing their survival and health once they arrive at the supermarket or restaurant.

Fish handling ensures that the farm's tank space is being used in the most efficient way possible. There is a balance between maintaining maximum stocking densities in tanks on the farm and reducing the number of times that fish are handled. Each operator will come up with a method that makes sense for them and their farm, but it should be noted that keeping tanks full of similarly sized fish can be difficult at times.

Chapter 7: Operational Considerations

A key piece of fish handling is keeping track of all the fish in the system and understanding the production goals and desired stocking densities. Fish leaving the system to go to market need to be replaced by fish coming in. This can only be done efficiently if there is a good understanding of how many fish are in each system, their average size, and the time left to harvest. Stocking frequency, stocking size, and stocking density will all vary based on the species. Also the number of gradings, and total time in the farm will vary. Planning when and how often fish are handled requires considering each of these factors and it is often the job of the production manager to make these determinations based on the information he has available. New software can now help a farm manager keep track of all the fish in the farm and help with planning where and when fish should be moved. An Excel spreadsheet also works as long as it is frequently updated.

The keys to good fish handling practices are to keep the number of times the fish are handled as low as possible, reduce stress on the fish, use the right equipment, keep track of fish movements, and properly plan all movements. Following each of these steps will make production goals easier to achieve and ensure the farm is operating efficiently.

System Maintenance & Corrosion

Regular system maintenance is needed to keep equipment in working order and prevent catastrophic failures. A water treatment step or an entire system may need to be taken offline to perform some maintenance tasks. As a result, it is best to

Chapter 7: Operational Considerations

plan these activities when fish biomass is low or at the very least during a break between feeding times. Often a system can be shut down for a brief period of time and the emergency oxygen system can maintain enough oxygen in the water. Better yet, would be to have designed the system to have redundant or parallel treatment systems so that one half can be taken offline periodically for maintenance.

Drum filters, pumps, blowers, oxygen generators, heat pumps, UV filters and sensors are all examples of equipment that require regular maintenance. Equipment manuals will outline what regular maintenance is required for each piece. The maintenance schedules of each piece of equipment should be easily accessible to all farm staff and a regularly updated calendar should be publicly posted. Maintenance most often involves replacing parts that wear down frequently, resupplying consumables, or cleaning air filters. Consumables or commonly replaced parts should be kept on site to allow for quick replacement.

It is also common to keep extra backup pumps or blowers on site. The advantage is that a failed piece of equipment can be replaced within hours instead of waiting weeks for an order to be processed and shipped. The disadvantage is this requires more upfront investment in equipment. An experienced operator will have an idea of what backup equipment should be kept on site and what pieces are unlikely to fail and do not require a backup. Another option is to source equipment from suppliers who are close by so that they can quickly provide replacement parts in case a rapid repair is needed. Similarly,

Chapter 7: Operational Considerations

equipment manufacturer representatives who are close by can quickly respond to a service call and repair broken equipment. This is a huge advantage and finding local suppliers should not be overlooked during the purchasing process.

A final piece of system maintenance is managing corrosion of equipment. Anyone who works around water knows the power of corrosion, although saltwater is the most corrosive, freshwater systems will also corrode metals over time. Corrosion is electrochemical in nature and it requires an electrolyte solution for electrons to move around, the more electrolytes that are in the water the more easily electrochemical reactions can occur. All aquaculture system waters are electrolyte solutions full of sodium, chloride, calcium, magnesium, and other ions. The general corrosivity of the water is dependent on many factors including salinity, temperature, pH, alkalinity, and hardness. Galvanic corrosion involves an anode losing electrons and a cathode gaining electrons due to a redox reaction, the loss of electrons resulting in corrosion of the anode. This leads to rust, pitting, and loss of material, which can eventually result in equipment failure.

Corrosion can be prevented first and foremost by choosing the correct metal alloys for your system. A saltwater system should have primarily 316 SS or plastic equipment; whereas coated cast iron and 304 SS can be used in freshwater systems. However, if the freshwater system has a high level of TDS and conductivity (strong electrolyte solution) then 316 SS equipment may need to be considered. Generally 316 SS and other corrosive resistant metals cost more, but the savings in

Chapter 7: Operational Considerations

reduced equipment replacement is worth the upfront investment. Once corrosion starts it can be difficult to stop, but there are coatings available that can be applied to rust spots to slow the corrosion down.

Another method to prevent corrosion is the use of sacrificial anodes; in boats or industrial plants zinc and aluminum are most commonly used. However, for aquaculture both zinc and aluminum are potentially toxic to fish and their dissolution into the water is not desirable. A magnesium anode could be used in some cases but it should be regularly checked to see how quickly it dissolves. The use of a sacrificial anode also requires all equipment to be electrically tied to that anode through conductive cables.

Equipment maintenance keeps a system operating smoothly and predictably. Failures will still occur, but being a good operator is all about managing and preventing risk. Keeping spare parts on hand and adhering to regular maintenance schedules are two key pieces to making sure the farm is always operating smoothly.

Purging

Purging is the act of placing fish into clean water to eliminate off-flavors, often described as a musty or muddy flavor. The off-flavor in the fishes flesh is often caused by the presence of geosmin, an organic byproduct of some bacteria. Thus the goal of purging is to eliminate the presence of geosmin and bacteria in the water. Geosmin is detectable by the human nose at

Chapter 7: Operational Considerations

extremely low concentrations (5 parts per trillion) and can ruin the flavor of an entire fillet. The off flavor issue is prevalent in most RAS because of the large populations of bacteria in the systems biofilters. Fortunately, if fish are placed in a clean water source without geosmin for approximately one week the off-flavor will disappear from the flesh. Every RAS operation has different levels of geosmin in the water and will require different purge durations for the fish in that system. Often times the flavor is not detectable in every fish or even every harvest, but it is a good practice to purge regardless since a single bad fish can ruin the experience for a customer and damage a farm's reputation.

The purge should take place for a standard length of time for all fish harvests (typically 4 - 7 days) and advance planning is needed to ensure the fish are fully purged before they are harvested. Often the replacement water that is entering the farm directly from the water source first passes through the purge system before being sent onward to the rest of the systems. This allows the bacteria-free, pre-filtered source water to be first put to use in the purge system before again being used in the main grow-out systems. General system water should never be used in the purge system since it may contain geosmin. In the purge tanks the fish are not fed, as a result the ammonia and oxygen loads are significantly reduced. Aeration may be required to keep oxygen levels high and carbon dioxide low, but additional water treatment is typically not needed in the purge system. After fish are harvested from the purge system, it can be drained and dried to prevent the growth of bacteria on the tanks walls, floor, and any piping. The

Chapter 7: Operational Considerations

use of a purge tank means fish must be harvested in batches or at least partial batches. Multiple purge tanks may be needed based on the harvest schedule. Fish should not be kept in the purge system for an excessive amount of time, as they will begin loosing weight since they are not being fed, this reduces the farm's yields. Future research is looking into how geosmin can be eliminated from the water through other methods, and many farmers have their own anecdotal evidence for how purging can be done more quickly. In the meantime it is best to stick with a consistent purge schedule and expect that this will be a part of your farm's weekly operations.

Feeding

Good farm management is good feed management. A good operator aims to maximize the amount of feed ingested by the fish each day, thus increasing the fish's growth rate. The whole farm operation revolves around making sure fish are happy and healthy so that they can eat and grow. As previously discussed, feed is the highest operating cost of any fish farm. We also have previously discussed the various feed equipment choices. However, just as important as the equipment are the type of feed, the amount fed, and the timing of the feedings. Controlling these three factors will have significant impact on the FCR and subsequently farm profitability.

Feed comes in a variety of pellet sizes, smaller sizes for smaller fish, larger for fully grown fish. A typical RAS operation may require three pellet sizes on hand at any one time; each size is for a different life stage of the fish. Different fish have different

Chapter 7: Operational Considerations

dietary needs and species-specific diet formulations are available for commonly cultured species such as Atlantic salmon, rainbow trout, and tilapia. There are now also RAS specific feed formulations that include additional binders that lead to increased water quality by helping the fecal matter stay together and decreasing the leaching of fats and proteins into the water from uneaten feed. There are two multinational fish feed companies, Skretting and EWOS. Both have a number of product lines at different price points. Shipping costs are often a major consideration when considering a feed provider. There are also smaller local companies that may also be a good option. Many farmers test different feeds to decide which they feel is best for their fish and system. Continued innovation in the feed formulation industry will lead to lower prices, reduced fish meal needs, and improvement in RAS water quality.

The amount fed per day will also depend on the life stage of the fish, generally speaking younger fish require more feed per day as a percentage of body weight. This ranges from 1.5% Kg of feed per Kg of fish per day for mature fish to 5% for fast growing juvenile fish. These percentages are daily goals that the operator wants to achieve. A variety of factors may affect whether or not those goals are achieved each day.

Disease outbreaks or adverse water quality conditions will reduce the appetite of the fish and less feed will be consumed per day. Observing how fish feed day after day is one of the easiest ways to discern changes in fish health. Less vigorous than average feeding behavior indicates that something is amiss. Daily observations of feeding are an important part of

Chapter 7: Operational Considerations

the farm routine, and it is important to train all operators how to interpret feeding behavior. In addition to the behavior of the fish, one can also monitor the amount of feed left over after a feeding event. In large net pen operations, underwater cameras are often used to monitor feed at the bottom of the cage, this can also be implemented in large tanks. Another method is to monitor the water flow returning to the drum filters from the bottom drain of the tanks to identify and quantify the amount of left over feed. A screen can temporarily be placed in the flow to capture uneaten food, providing an easy visual of what is left over. Leftover feed is a direct hit to the farm's bottom line and it puts a load on the filtration equipment, but it is expected in any farm.

The last factor to control for feeding is how often pellets are distributed during the day. Different farmers and different species require different schedules. Feed can be fed nearly continuously, with automatic or demand feeders, or it can be fed in regular batches a few times per day. Most fish should only be fed during the daytime or whenever lights are on in the building. A regular schedule allows for easier day-to-day comparisons of feeding behavior.

A good operation should carefully track feed use on a daily basis. Feed tracking is an important part of being able to optimize system operation and increase profits. Different feed methods require different feed tracking systems; find one that works well for your operation. Overall, take the time to feed the fish in a predictable, measureable way. RAS operators can take cues from how net pen operators manage their feed. Feeding

Chapter 7: Operational Considerations

is a detail-oriented job that is continuously tweaked to increase FCR. Experience is required to properly feed fish, but becoming great at it is one the keys to a successful farm.

Biosecurity

Maintaining biosecurity is a continuous fight against complacency. Operational protocols must be put in place and fairly enforced by managers. Many farms fail to take the proper steps to maintain biosecurity, and unfortunately they only realize their mistake when a massive disease outbreak completely derails their farm, bringing revenue to a grinding halt and ultimately bankrupting the business. Don't make this mistake.

There are a few different aspects to biosecurity. First there is the disinfection step during the water treatment process. As previously discussed, this is best accomplished with a UV system, the UV system works to continuously knock back the population of potential disease causing bacteria, viruses, and parasites. Suppressing the population of bad actors and reducing the stress of the fish at all steps will help to keep diseases in check. In order to ensure that proper disinfection is occurring you need to perform regular check ups on the UV equipment as part of the routine maintenance.

The next step is creating a perimeter defense around the farm; this prevents potential diseases from entering your farm (Figure 16). This is accomplished by making sure anything that enters the farm, including people, is thoroughly disinfected. In order

Chapter 7: Operational Considerations

to accomplish this the points of entry and exit into the farm should be extremely limited. All personal and equipment can enter and exit through a single door that has an anteroom where clothes and shoes can be changed or disinfected. Inside of this room there can be a rack for where shoes can be changed out for sterile boots, and hands can be disinfected with sanitizer. Also there is often a sign-in and sign-out sheet to keep track of who is visiting the farm, this can help determine the source if an outbreak were to occur. A footbath, consisting of a shallow basin full of iodine disinfectant, should be placed at the entry door to ensure all feet are washed before entering the farm. Many farms do not allow outside guests inside of the building, this is because they do not know if guests have previously visited other fish farms and have potentially been exposed to a disease. If allowing guests on site you can provide them with full body Tyvek suits, gloves, and boots. Guest visits inside of the farm can be cut down by having a viewing area outside of the main farm that has windows looking into a farm, this way prospective clients, investors, etc. can see the farm without having to pass through biosecurity protection.

Chapter 7: Operational Considerations

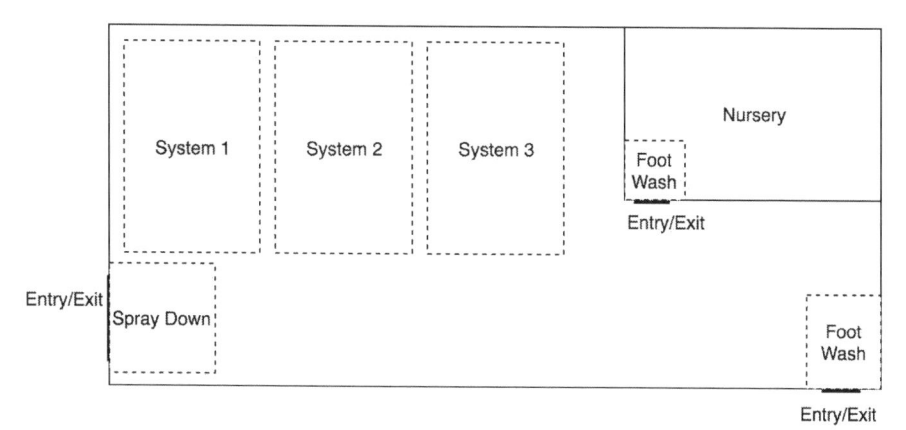

Figure 16. Example of biosecurity borders within a farm. Solid lines represent solid barriers such as walls, and dotted lines represent informal borders for wash downs.

It is likely there will also need to be a larger entry and exit door into the farm, which is used less frequently for deliveries of feed and equipment. Equipment should also be sprayed down with disinfectant before entering the farm, or if it is still in the packaging it can be assumed to be disease free. Chlorine can be used to disinfect equipment, but under no circumstances should chlorine come in contact with the system water. It can have extreme adverse affects on the health of both the fish and the biofilter. Rinse any equipment with water after it has been disinfected. And it should go without saying that if any pests are detected at the farm, such as flies, spiders, rats, or birds these should be removed immediately as they can also be a possible disease vector.

After securing all entry and exit points for the farm the next step is to maintain biosecurity between systems. Keeping systems isolated from each other is the easiest way to ensure an outbreak does not affect the entire farm. Isolation between

Chapter 7: Operational Considerations

systems can be maintained in a couple of ways. The easiest way is to physically isolate systems by constructing walls between them, this is not always practical but should be done if possible. Isolating the nursery system from the grow-out systems often makes the most sense and is recommended. This way potential diseases that are present in newly arrived fish are always isolated from the main population of existing fish on the farm. Next, even if systems are located in the same room water should never be exchanged between systems. Also equipment that regular comes in contact with the water such as nets, fish pumps, graders etc. should either only be used in one system at all times or be disinfected when moved between systems. This can be cumbersome and labor intensive at times, but it is a best practice to ensure biosecurity. One of purposes of having multiple systems in one farm is to create redundancy. Take the time to isolate systems properly and your chances of catastrophic mortalities from disease are greatly reduced.

Next is to make sure all fish entering the system are being sourced from a biosecure facility that is certified disease free. Fish hatcheries have an even greater incentive to keep their facilities disease free since there reputation is spoiled if any customer receives diseased fish. Nonetheless, deal only with reputable egg and fingerling dealers; in this case the cheapest source is probably not the best option. When fish or eggs first enter the farm they should be kept on a separate system for at least a few days or possibly longer. The fish should be regularly observed for any potential symptoms before being placed in a system with existing fish.

Chapter 7: Operational Considerations

Making the effort to maintain a biosecure facility is worth it for the peace of mind. The challenge is to make sure standards are not relaxed after running a farm for years without a problem. Biosecurity is just another part of being a good operator.

Disease Management & Treatment

If the fish are showing disease symptoms it is not too late to begin implementing measures to potentially contain the outbreak. The first step is to identify the disease and provide aggressive treatment, this may mean changing the fishes diet, or adding supplements to the water. Bacteria, viruses, fungi, or parasites can cause common fish diseases. A veterinarian may be able to provide a rapid diagnosis if provided with a diseased specimen. It is good to know your local fish veterinarian before your first outbreak occurs.

In addition to providing aggressive treatment, the suspected system should be isolated from all other systems, a good practice is to put up temporary physical barriers that staff cannot pass through with out first going through a thorough cleaning. Feed loads should be reduced on the diseased system, extra oxygen should be added to the water if possible, and additional water should be exchanged. All of these measures help to reduce stress on the affected fish. Other systems in the farm should be closely monitored for signs of disease.

In some cases the best course of action may be to immediately euthanize all fish and drain the system. The system should then be completely disinfected by letting all parts of the system sit

Chapter 7: Operational Considerations

dry for a number of days and spraying everything down with a bleach mixture. This is an extreme measure but many farmers have had do it after repeated failures to completely remove a disease from their farm.

Look for disease management resources and books that can help diagnosis possible diseases and provide treatment guidelines. Most RAS operators are not interested in using antibiotics as it hurts the marketability of their product. If you choose to use antibiotics or any other disease treatments make sure to carefully follow the manufacturers instructions and reach out to company representatives if you have any questions.

Staffing

Like any business, hiring competent, reliable staff is both the key to success, and one of greatest challenges. It takes a special skill set to identify good candidates and train them into great employees. Depending on the size of the farm there will be different staffing needs, however 24-hour on-site staff is needed for any farm. The night shift on a small farm can still sleep on site, but they need to be able to respond to an alarm immediately at any time.

Larger farms will require one or more farm managers that are responsible for day-to-day decision-making and ongoing production planning. The farm manager becomes the point person for all farm related problems. Finding a good farm manager or becoming one yourself is one of keys to successfully operating a profitable RAS. The farm manager

Chapter 7: Operational Considerations

needs to have a stake in the success of the farm and be willing to undertake the task of running the farm so that it performs and produces high quality fish at the right price. Responsible managers are hard to find, and it may be necessary to train them in house. The best managers are good at understanding fish, equipment troubleshooting, production planning, and personnel management. It is a rare person who can excel at everything and having multiple managers with different areas of responsibility may be more effective.

Underneath the farm manager may be other production staff that perform day-to-day tasks such as feeding, fish handling, water testing, etc. A lot of the work at a farm would be considered physical labor, and as a result the jobs are not usually high salary. However, a lot of different skills are needed to be a good RAS employee. First, they need to have a good understanding of fish biology and basic farming techniques. Second, they need to understand how to operate and maintain equipment. And most importantly they need to be reliable, attentive, and flexible. Often the production staff has a high turnover since the pay is low and the tasks are repetitive. Paying good wages and goal-based bonuses are methods for reducing staff turnover.

In addition to the on-site staff there may be a need for additional staff on the accounting and administrative side of the business. In small farms the owner typically handles this. Larger farms though may require staff for administrative tasks such as accounting, sales, and shipping. Keeping things organized on the business side of the farm is important and cannot be

Chapter 7: Operational Considerations

neglected, fortunately it is easier to find staff that can perform the administrative tasks as the same skills needed in other fields can be applied to the aquaculture business.

Chapter 8: Other Topics

Regulations & Permitting

Regulations that will affect a RAS take many forms. There are business regulations, construction regulations, and aquaculture regulations all of which may be enforced at a federal, state, county, and city level. Navigating these regulations can require a PhD in paperwork, semantics, and patience. Maintaining a positive attitude and finding allies within the local government is the best way to keep things moving along, especially if you hit any roadblocks. Jurisdictions that have previously dealt with permitting aquaculture operations will most likely be easier to navigate and you will be more likely to receive approval for the project in a timely manner. There can be a lack of accountability with certain government organizations in providing timely responses and approvals to permit requests. Similarly, unfamiliar projects tend to get put on the slow track with no officials willing to take responsibility for being the individual who says "yes" to the project. Unfortunately this creates an atmosphere where innovative projects, like a RAS, are stalled while waiting for approval. It is always best to take care of permitting and regulation compliance during the early stages of planning as suggested in the design steps chapter of this book.

If you are located in the United States there is often a contact person for each state at either the fish and game department or agriculture department that can help with understanding what permits are necessary for your operation. A common set of permits that would be required is a fish handling permit, wastewater disposal permit, business license, and construction permits. This is not a comprehensive list and each jurisdiction will vary. Talk to fellow farmers in the area where you are planning to build your operation, many have had to jump through the same hoops.

Construction & Engineering Partners

Choosing good project partners is one of the keys to success with RAS. Unfortunately, when money gets involved, a lot of good people start behaving badly. This is where, as mentioned before, having good contracts in place ensures good behavior and creates clear expectations. Finding trusted partners is often accomplished the old fashioned way through word of mouth. The aquaculture industry is a small world, populated by only a handful of companies, and of these companies only a few have had success in RAS projects. Look carefully at the project profiles of any companies you choose to work with and check references. This takes effort and time, but it is the easiest way to vet potential partners. At the end of the day the RAS owner is the client, and who you choose to work with is who you are going to be spending money with. As the customer you want to make sure you get the best value from your engineering and construction team.

In general, construction is a very inefficient process, in fact in the US it has reduced in efficiency over the past fifty years, less work gets done per man hour today then it did in the 1960's. This is due to a variety of factors, primarily the inability to

Chapter 8: Other Topics

automate tasks since the work is all customized and complex. This inefficiency can be very frustrating for the uninitiated. Setbacks and confusion commonly occur, and getting everyone on the same page is one of the biggest struggles of any construction project. Additionally a RAS is not something that very many people have built before and most likely almost all of the general laborers and construction managers on your project will be unfamiliar with how a farm is suppose to look. What all this means is there will need to be constant meetings and communication between all project partners: owners, engineers, general contractor and subcontractors. All of this time and talking about the same things over and over again is normal and to be expected.

Of course before construction can even begin there will also have to be a lot of meetings and discussions with your engineering team. Working with an experienced engineering partner is perhaps the greatest favor your can do yourself. A proven track record of successful projects that are still in operation is the greatest indicator of the quality of work you can expect from an engineering firm. Look for partners who are transparent and are willing to openly and critically discuss any projects that they have been a part of in the past that may not have worked out.

Every engineer has there own methods and theories for how water treatment can best be accomplished, these differences between engineers is beginning to disappear as the whole industry realizes what water treatment processes are scalable. Nonetheless it is important to understand what methods the

Chapter 8: Other Topics

engineers suggest and to be able to ask critical questions about why this method has been chosen over other alternatives. It is best to avoid any novel technological processes that have not been previously tested on a commercial scale. Engineers also have a tendency to focus solely on the water treatment process and pay less attention to operational considerations such as fish handling and feeding. The engineer should assist in putting all of the components of the farm together including buildings, electrical, piping, controls, water supply, and wastewater treatment. All of these pieces need to work in concert, and getting everything right often involves bringing all project partners, including equipment suppliers and contractors, together to discuss and work through sticky problems; this is best coordinated by a competent engineering team.

In summary, pick the right partners. Look for good recommendations, a proven track record of successful projects, and a strong willingness to work collaboratively. Avoid partners who do not stand behind their past projects and avoid answering difficult questions. Expect decision making to take a lot of time, and for the design and construction process to move along slowly.

Fundraising

A lot of first time RAS owners come into the industry from the business world, but for those who are not as experienced in fundraising there can be a steep learning curve. The first thing to know is that fundraising takes a lot of time. Since there are so few successful RAS operations to point towards, convincing

Chapter 8: Other Topics

investors to back your project will be a huge uphill battle. First, banks will not be interested in providing debt financing since there is not a proven track record of business successes. Equity financing, though possible to obtain, will take months or perhaps even years of back and forth talks with potential investors. The key here is to cast a wide net and use personal contacts to talk with as many potential investors as possible.

Receiving venture capital money is extremely difficult, it may seem easy when you hear about all these Silicon Valley companies raising millions every month, but the reality is far bleaker. The goal of a venture capitalist is to invest in projects that are often higher risk but also higher reward. Unfortunately, most RAS projects do not fit this mold, they don't have the potential like technology startups to grow a thousand fold in a couple of years. Instead they return a steady income stream month after month and year after year. Investing in a farm is like investing in a dividend stock. Venture capitalists have a bad habit of never saying yes or no, they want to keep as many potential deals hanging in the air at all times in case one potential investment suddenly becomes hot. This can be extremely frustrating as a fundraiser, the decision seems to be perpetually delayed, excuses that one more thing needs to be checked into, or one more person needs to consulted is a common delaying tactic. If a potential funder seems to be delaying, the best thing is to leave the door open, but to not hold your breath for any kind of deal and instead look for money elsewhere. The CEO of a RAS operation may spend as much as 80% of their time just trying to bring funding together, this leaves less time for considering design changes, seeking

Chapter 8: Other Topics

better bargains on equipment and construction, or performing market research.

There are a couple steps you can take to give yourself an edge in finding venture capital. First have something to show investors; this can mean having a small pilot system that was funded with your own money, or even having some fish on hand to show off the quality. This shows your commitment to the project, and it proves that you have already invested time and money. It also helps tell the story of how your business is going to be successful. The next biggest advantage you can give yourself is to have potential customers already signed on to purchase agreements. This can be in the form of written statements that say they would be interested in buying the farm's product or actual signed purchase agreements with guaranteed sale prices. This shows the investors that there is a guaranteed market for your product. Of course it is getting a bit ahead of yourself to begin signing up customers before you even have a farm built, but it is a common tactic in many other industries. There are countless resources available to help you craft your pitch, find your marketing angles, and tell a compelling story that will entice investors to trust their money to you.

At the same time don't hide anything from potential investors, you are just lying to your investors and yourself if you are not completely transparent about possible pitfalls of the project. Venture capitalists are paid a lot of money to figure out what is going on behind every door and they will thoroughly vet the finances of any project presented to them. It will save yourself

Chapter 8: Other Topics

and investors time if you are just honest from the beginning with your business plan and cost/sales forecasts. The fact is if you have to hide something, then the business prospects of the operation are not good and you should spend your time thinking of new ideas. Some of the best business advice available is if you have a new idea you should do everything to kill that idea, think of all the possible reasons why it is horrible idea, and ask those you trust to look at your idea critically and think of all the ways it sucks. If at the end your idea has withstood all of this criticism, then it must be a good idea and is worth pursuing.

An easier route then venture capital financing is to self-finance a project, or raise money from friends and associates. This removes a lot of headaches from the business side of the farm, and ultimately it leads to a higher chance of success. First you will not have to waste valuable time on fundraising. Second, you will not be accountable to a board of investors or any other outside pressures that may force you to compromise or make ill-advised decisions. Also when you are spending your own money or the money of your friends you are much more likely to spend it wisely. It is always easy to spend other peoples money, but with your own money you make smarter decisions and spend more time looking for cost effective solutions to building and operating the farm. These decisions are the ones that lead to a successful farm.

It is important to keep in mind with however your decide to raise money that you will want to have at least 24 months of operating cash on hand when the farm is first started. It can

Chapter 8: Other Topics

take a long time for cash flow to get started and the first fish often take at least one year to grow before harvest. Also it is likely there will be some initial setbacks in achieving predicted production levels and having the wiggle room to make a couple mistakes will keep you from running out of cash before you can start making money.

When budgeting your project, crafting the business plan, and raising money, you should assume the worst-case scenario for the market price of the fish, the capital cost of the farm, and the cost of production. Also assume that construction will take substantially longer than you anticipate. A common rule of thumb among aquaculturists is to assume twice the costs, and twice the construction time when planning the initial build out of the farm. The market price of your fish will also most likely not be what you would like it to be, especially if you are flooding the market with a novel species that no one has previously tested.

In summary, take the time to form an honest business plan, understand the ins and outs of your market, figure out whom your end customer is and do research on their needs. Be pessimistic while planning and raise more money then you need. Finally, realize it will take time and immense effort to raise money to fund a big project.

Aquaponic Systems

Aquaponic systems grow both fish and plants on a single system where the fish water fertilizes the plants, which in turn

Chapter 8: Other Topics

helps remove some of nitrogen and other nutrients from the water. Aquaponics has grown in popularity in recent years and many hobbyists have built small backyard systems to grow fish and plants. Commercial scale success though has been hard to come by. Over the past twenty years many aquaponic operations have tried to get off the ground, some well funded by venture capital, and none have been financially successful in the long-term. Some farms have claimed to be financially feasible and offer to sell design help or workshop classes to teach prospective farmers, but often the selling of classes is actually their main revenue source.

The failure of most farms is the inability to produce either fish or plants at a competitive price point. Unfortunately with two products you need twice as much knowledge, and you end up having twice as many headaches. There is a reason why orange farmers only grow oranges and chicken farmers only grow chickens, they master one product and focus on producing that product for less. The challenge in trying to grow two products that are linked is that neither can be optimized. Fish and plants require different environments to maximize growth, and the strength of indoor, closed systems is the ability to control all environmental factors to increase production. Instead of optimizing, you are forced to compromise between two competing sets of ideal conditions. This causes you to become uncompetitive in both markets by making the cost of production too high for both the fish and the plants. It is important to remember that you are competing in a commodity market against some very large-scale producers. Greens and other vegetables are grown in fields on huge scales that

Chapter 8: Other Topics

significantly reduce the cost of production and consumers are only willing to pay so much of premium for food grown in a novel way. The same story is true for fish.

The correct way to approach aquaponics is to focus on either fish or plant production and consider the other half an ancillary revenue source and thus of lesser priority. Optimizing for plant growth means treating the fish as a source of fertilizer for the plants. In this scenario the plant beds will take up considerably more square footage than the fish system. Many hydroponic growers are now using tiered stacks of plant beds that use specialized LEDs to provide light to the plants, thus reducing the building square footage they need. It is also possible to use these kinds of systems in aquaponic operations.

The other option is to build a RAS for fish, focusing only on how the fish should be grown. Then a greenhouse or other ancillary building can be added and either the daily replacement water or loop of system water can be circulated through the plant system. In this scenario the square footage of the plant beds may be similar to the square footage of the fish system.

The benefits of aquaponics are the ability to diversify into two crops, creating two independent revenue streams that are not dependent on the same price fluctuations or other market winds. The drawbacks are the increased complexity of the system and inability to simultaneously optimize conditions in both systems. A couple examples of this optimization problem: first is the temperature of each system, for most leafy green the ideal water temperature is around 23 °C while the ideal growing temperature for warm water fish such as tilapia is around 28 °C. Another example is that the presence of essential nutrients such as iron and phosphorous is required for adequate plant growth, while these same nutrients can be harmful to the fish if concentrations are too high. Of course there are creative solutions to mitigate these mismatches, but those solutions require time, energy, manpower, and most importantly, money.

As some point in the near future the code to operating a successful aquaponic facility may be cracked, but in the meantime it is best to approach the business with caution. The same planning steps outlined in this book for RAS should also be applied to each aspect of an aquaponic operation. Despite its drawbacks it is still a great way for hobbyists to grow some food while learning how to grow plants and fish in a controlled system.

Hydroponic Systems

A hydroponic system is an aquaponic system without the fish, and as a result it looks a lot different. No water treatment is needed in a hydroponic system, instead water is circulated with a pump between grow beds and nutrients are added to the water to fertilize the plants. Only large solids may need to be filtered occasionally as leaf or roof parts can fall into the water. The challenges in designing a good hydroponic farm are all related to how plants can be handled and harvested with minimal labor. Large hydroponic systems move rafts of plants

Chapter 8: Other Topics

around with channels and robotic arms that can pick up and move rafts into adjacent ponds or onto processing equipment.

The advantages of hydroponic systems over traditional agriculture are the same as RAS for fish culture. Less water use, more control of the environment, and the capability to grow crops anywhere. Generally speaking a hydroponic system would not be classified as a RAS since it is not used to grow aquatic species. However, a lot of the same people who are interested in RAS are interested in hydroponics because of their shared advantages. New hydroponic businesses are starting to use stacked trays inside of warehouses that are lit with LEDs. These companies site their farms near large urban areas and sell their product for a premium price to upscale grocery stores. RAS companies can look towards large-scale hydroponic growers to understand how they can replicate their successes and use similar methods to sell and market their seafood products.

Polyculture Systems

Polyculture is the process of growing two or more species of fish, invertebrates, or plants in a single system. Aquaponics is a form of polyculture, and scientists have been exploring the use of more creative combinations of plants and animals to create zero-waste systems. Examples include using shrimp or worms to clean up fish solids by placing them in a series of tanks directly after the fish tanks, or using seaweed to remove nitrogen, carbon dioxide, and phosphorous. There are almost infinite combinations to try in both marine and freshwater

Chapter 8: Other Topics

systems. For the moment though, polyculture in RAS is still experimental and there are currently no profitable businesses using polyculture methods. Any polyculture system should be approached with great caution and planning, the same as an aquaponic system. Any new combination of species will need to be pilot tested first to gather the data needed to properly operate and build a full commercial system. Polyculture it most likely only going to be successful if all the species being grown are high value and each one somehow contributes significantly to reducing the water treatment equipment and energy needs of the entire system. The added complexity of managing multiple species must be counterbalanced by the subtraction of complexity somewhere else in the system. Otherwise it would just be better to grow each species in an independent, non-linked RAS.

Biofloc Systems

Biofloc systems are particularly popular with shrimp growers, and work has been done to use biofloc with tilapia systems. The benefits of biofloc are the elimination of the needs for a biofilter or significant solid filtration. A key element of biofloc systems is heavy aeration to keep the biofloc in suspension, this also serves to oxygenate the water while stripping carbon dioxide. Ultimately this leads to a very simple system where water does not need to be circulated. Furthermore, shrimp can feed on the biofloc, essentially recycling wasted food back into the system. Biofloc systems are often said to be capable of using zero water exchange during a single growing cycle.

Chapter 8: Other Topics

The biofloc itself is made up of heterotrophic bacteria that consume the excess carbon in the system from fecal matter and wasted food. In the process of consuming the carbon they incorporate the excess nitrogen in the system into their cells, so as the excess biofloc is removed from the system so is nitrogen. One of the keys to managing a biofloc system is to make sure that the mass of carbon and nitrogen entering the system via feed is added at the same ratio as it is present in bacterial cells. This generally is considered to be a ratio of fifteen parts carbon to one part nitrogen. Fish feed is often much higher in nitrogen than this ratio, and so as a result excess carbon must be added to the system in order to keep things in balance, this is most frequently done by the addition of a sugar such as molasses. Autotrophic nitrification will still occur in the system and water needs to be exchanged as nitrate levels begin to grow. Also attention needs to be given to phosphorous levels if water exchange is extremely low.

The challenge with any biofloc system is the management of the bacterial population to properly control the water quality. The reality is that managing a biofloc system requires a great deal of care and attention or the system will stop maintaining proper water quality. This is not a problem on a small-scale, particularly in research systems, but it is difficult on large scales. The issue is each biofloc system has an independent bacterial population that must be managed. It is common to see two side by side biofloc systems with very different water colors, ranging from green to shades of brown. Although these two systems may have been initially set up and maintained using the same methods, the different water color suggest the bacterial populations in each have diverged and they no longer can be treated the same. The challenge is to manage all of these slightly different systems at once and ensure they maintain the same high-level of water quality.

To date biofloc technology has not been demonstrated to be a commercially viable alternative to traditional RAS. Future research and development needs to be done to fully understand how biofloc systems can be scaled and operated predictably.

Hatchery Systems

Hatchery RAS can be operated economically and the business prospects are better than building a food production RAS. Hatchery systems have a double advantage, they support a relatively small biomass at a lower stocking density, and they produce a high-value product. This combines to make a low overhead, high revenue business model as compared to food production systems.

Typically a hatchery has a broodstock system where mature males and females are kept under ideal conditions until they either naturally spawn or are artificially spawned. The eggs are then kept in carefully controlled systems that often only require minimal filtration since no feed is going into the system, and thus no solids or ammonia is produced. For some species, such as salmon and trout, the eggs are sent to the farmer unhatched and the farmer completes the rest of the hatching and feeding process. For other species, such as tilapia, the eggs hatch and

Chapter 8: Other Topics

are fed their first meals at the hatchery. Again very little water treatment is required. Once large enough the fingerlings are shipped to customers with grow-out facilities, which could be ponds, sea cages, or RAS.

However, there are high barriers to entry for hatcheries. First a hatchery needs a high quality genetic stock of fish. Operating a hatchery requires selectively breeding the broodstock population to create high quality, fast growing eggs or fingerlings. Good genetics is required for supplying repeat customers. A distinctive line of high quality fish is created over time by selectively breeding many generations of fish. In order to accomplish this a high amount of capital is needed to start the program and sustain it over time while investing in high tech genetic testing. Furthermore, the customer base is very limited, as there are only so many farmers growing the species of fish that you are breeding, and gaining new customers could be difficult without an existing track record. Finally, often you are competing with big companies from all over the world. Eggs and even fingerlings can be shipped almost worldwide with airfreight, making competition particularly fierce. And there is a final risk, which is the eggs or fingerlings must be considered disease free at all times, a single disease outbreak could destroy a company's creditability and ultimately bankrupt it.

Shellfish Hatchery Systems

Many shellfish hatcheries are operated as flow-through systems, but some are operated as RAS especially if pH or

Chapter 8: Other Topics

temperature needs to be carefully managed. Shellfish hatcheries deal with a variety of larval invertebrates that are considerably smaller than larval fish. These microscopic larvae are extremely fragile and require near perfect water quality in order to survive their first few weeks of life. Traditionally, in order to achieve this high level of water quality shellfish hatcheries were operated as flow through systems and were located near a high quality water source, such as a remote tidally flushed bay. The drawback of this is you are dependent on the source water to maintain water quality in the hatchery. If a storm increases sedimentation, or a toxic algae bloom moves into the area you have to pump the less than ideal water into your hatchery to keep the shellfish alive. You can always treat the incoming water with UV and solids filtration, but if the water quality is regularly compromised it makes sense to begin recirculating and treating water within the farm.

Fortunately, just as with a fish hatchery, biological loads in a shellfish hatchery are almost non-existent given the small size of the larvae, often they do not feed for the first few weeks of life and when they do begin feeding it is on algae. This means solids filtration, biofiltration, and oxygenation systems can be relatively small and equipment costs are low. A heat pump to maintain temperature would most likely be needed, as would a small control system to keep all parameters in line. The recirculating system gives a hatchery independence from its water source, which is one of the main advantages of any RAS. Shellfish hatcheries also almost always operate their own microalgae production systems to provide feed for the growing

Chapter 8: Other Topics

shellfish larvae. See the next section to learn more about these systems.

Algae Systems

Algae systems are used to grow food for larval fish and shellfish, as well as for biofuel production or food supplements. Algae production systems are often operated as RAS, but the biological needs of algae are very different from fish and as a result an algae RAS does not look like a fish RAS. The systems provide the correct environment for growing algae, this means lots of light, carbon dioxide, and essential nutrients. Algae systems also vary based on if they are growing macroalgae, also known as seaweed, or microalgae.

Microalgae are grown in both continuous culture systems and batch systems. In a continuous culture system new water and nutrients continuously enter the system at a slow rate and algae rich water leaves the system at the same rate. Carbon dioxide or air is often bubbled in and natural sunlight or fluorescent/LED bulbs provide light. Batch systems by contrast consist of inoculating a small container with algae and then increasing the size of container/tank as the algae grows. Eventually the algae is moved into a large tank that is slowly filled with water until it is full. At that point all the algae from the tank is harvested, and a small amount is saved to inoculate the next batch. Microalgae grow best when kept at a consistent stocking density, which can be carefully controlled in both types of systems.

Chapter 8: Other Topics

Macroalgae can also be grown in RAS, but to date it has not been shown to be economical. The uses for macroalgae are more limited and thus the sales price does not justify the costs of electricity and labor to operate the system. Creative marketing and LED light technology may make macroalgae farms viable in the near future.

The two main challenges in algae culture are first maintaining a sterile culture that is not overrun by bacteria, and second providing ample light for photosynthesis. Maintaining a sterile culture involves making sure the inoculate is sterile when starting a new batch and keeping the growing conditions just right to ensure there are no bacterial outbreaks. If an outbreak occurs the bacteria can steal the nutrients away from the microalgae and cause the population to crash. Ample lighting is simply a matter of siting the system in a location where there is a lot of natural sunlight or providing artificial light with high-pressure sodium or LED bulbs. In the past using solely artificial light to grow the microalgae has proven to be prohibitively expensive, but new LED technologies are beginning to lower production costs.

In both macroalgae and microalgae culture carbon dioxide can be bubbled into the water to increase the available inorganic carbon for photosynthesis, often leading to higher growth rates. Oxygen is produced during the photosynthetic process, but oxygen is also consumed at night by respiration, and may need to be bubbled in if concentrations are too low. Algae culture is an entire field of its own, and additional resources can

Chapter 8: Other Topics

be found if you are interested in growing algae for fish food or profits.

Post-Smolt Salmon Systems

Land-based Atlantic salmon post-smolt systems have been rapidly growing in popularity in recent years. Salmon smolt production has almost always been done exclusively on land in either flow-through or RAS systems, taking an average of 12 months to go from egg to smolt. At the smolt stage the juvenile salmon are ready to transition from freshwater to seawater. However, new methods have allowed farmers to begin the smolting process of maturing the juvenile salmon without salt water by instead manipulating the water temperature and photoperiod. This has led to holding the post-smolts in land-based systems for another 6 to 12 months. The popularity of this process has led to the construction of large RAS systems (>1000 MT/year). The high capital and operating costs of the post-smolt systems is economically justified because it allows the salmon farming companies to get more production out of their existing sea cages. Given the expense and difficulty of permitting additional cages, it has proven profitable to instead look for ways to improve the production rates of existing cages. Post-smolt systems allow cages to be harvested more quickly; larger stocking sizes means the fish can be harvested after one year in a cage instead of two years. This effectively doubles the production of each cage. The trade off is the need for the large RAS systems. The advent of these systems has led to additional technological and management

Chapter 8: Other Topics

innovations that will eventually spill over into other forms of RAS.

Currently, growing Atlantic salmon to harvest size in land-based systems is difficult given the amount of water treatment needed to support thousands of 5-10 kg fish. Similar to a hatchery system, a post-smolt system is not growing food fish but instead is producing a higher value product that can than be grown out in a different system that can be operated more economically. Smolt and post-smolts are worth more per pound then consumers would be willing to pay for on a per kilogram basis for food, but they are valuable to companies who can continue growing the fish to a larger size where the price per kilogram will go down. These kind of profitable RAS are what is needed to bridge the gap towards eventually building RAS that can grow any fish species to market size profitably, which currently is something that is only possible for niche species.

Seafood Holding Systems

Seafood holding systems are often set up as RAS because they are frequently not located near any reliable water sources. Also many systems are saltwater, and are located nowhere near an ocean. The purpose of the system is to receive catch from boats or farmers and hold the fish or shellfish before shipment to retail or other holding locations. These holding systems are often constructed to make handling of product very easy since the turnover of fish or shellfish in the system is high. This can make the design tricky, since the traditional use of large tanks

Chapter 8: Other Topics

and fixed pipes is not ideal. Instead systems will often utilize insulated, stackable totes and water will cascade down through the totes from a ceiling mounted pipe system that is out of the way of handling equipment and personal. Since labor is the largest cost in the seafood holding industry it makes sense to design systems around the use of forklifts and production lines. The seafood is often taken out of the holding system, packed on site, and then shipped out to clients in insulated boxes or totes.

One of the keys to seafood holding systems is an adequately sized chiller system that can hold the temperature around 4°C for coldwater species such as crab, lobster, flatfish, oysters, and clams. The advantage of the chilled system is the shellfish/fish being held in the cool water will have reduced metabolisms. This leads to less oxygen consumption, less ammonia production, and less movement. Because of this, biofiltration is often not needed in seafood holding systems and many utilize only disinfection, oxygenation, and simple solids removal equipment. Inline UV is most often used for disinfection, and oxygenation can be achieved with airstones or cascading water.

Stocking densities for seafood holding systems can be difficult to estimate since metabolism is significantly reduced and no feeding is occurring. What this does mean though is stocking densities can be extremely high. In some shellfish systems the product can be held at a ratio of one part shellfish to one part water, this is 10 times higher than the highest traditional RAS stocking densities. Given the lack of research and public information available on this subject a seafood handler should

Chapter 8: Other Topics

use trial and error to do their own research into what stocking densities make the most sense for how their system is operated. Seafood holding systems are often one-off constructions and their design is based on personal experience of operators or engineering companies. Pilot systems and experimentation may be needed to arrive at design guidelines for a large system.

Developing World RAS

There has been a great deal of interest in RAS technology for use in developing economies to generate income and create jobs. This sentiment is good, since all agriculture has a strong economic multiplier and is the kind of primary producing industry that forms the base of almost all world economies. However, the high capital costs and dependence on both reliable electricity and specialty equipment make it inappropriate for developing countries at the moment. It would instead be appropriate to promote the growth of other low-tech forms of aquaculture such as pond culture, flow-through systems, or in some cases primitive RAS. A primitive RAS can be constructed with the use of gravity solids settlers, a trickle biofilter, aeration for oxygen, and pumping with inexpensive plastic-body pumps. It is possible that in the very near future it will be possible to construct RAS in less developed economies and turn a profit. However, in the meantime more labor intensive forms of aquaculture that require more space are the most economical practices to encourage in developing economies.

Chapter 8: Other Topics

Hobbyist RAS

Most home built aquaculture systems are RAS. Rarely do individuals have access to enough water or space to do flow through, pond, or cage culture on their own property. Small aquaponic systems are currently the most popular form of hobbyist systems, this seems to be primarily due to the cool factor of operating your own closed loop systems that quickly yield a harvestable crop of plants. Commonly cultured species are trout in colder climates and tilapia in warmer climates. Goldfish are also commonly used to test a system out since they are cheap and easy to obtain, but they do not make good food fish. A hobby system cannot be operated as a profitable business, which would mean paying yourself a reasonable wage for the time you put in, but it can definitely be built inexpensively. There are a number of ways to recycle commonly available materials into a RAS.

Almost any tank, tub, bucket, or other vessel can be turned into a fish tank. Common cheap tank substitutes are livestock water troughs, IBC totes (typically used to ship food-safe goods such as cooking oil, glycerin, syrup etc.), and plastic storage bins. Using a single or a couple of large tanks makes things easier than trying to piece together a bunch of small tanks. Small PVC (1/2" – 2") can be purchased very cheaply from any hardware store. Cheap solid settlers can be made with buckets or with cloth bag filters that can be regularly washed out with a hose on the lawn. A cheap biofilter can be constructed using a gravel bed that water flows up or down through. The most important aspect of the biofilter is to have a high surface area, and many other inert materials can be used for an easy biofilter.

Chapter 8: Other Topics

Aeration can be achieved with either a cheap air pump from the aquarium store, or it is often times even easier to have the water cascade down through some substrate before entering the tank. A plastic pool pump is the best option for a cost effective way to circulate the water, these can be purchased online or sometimes at your local hardware store. Expect the pump to last from 2-5 years. A greenhouse is one of the easiest ways to shelter the system from the elements and maintain some heat in the system especially during the colder months for those in more northern latitudes.

A home system is a great way to practice problem solving and construction skills. Once the system is up and running new creative ways to improve the various waste treatment processes can be developed and tested. This is one of biggest advantages of a small system over a big commercial system, as it is often easy and cheap to switch out a few components and see if things improve. It is also easy to observe how the system responds to changes in stocking density and feed loads. Monitoring the water quality and the fish's health helps build intuition and understanding into how the fish affects the system and vice versa.

Robotics

Robots are gaining traction in all kinds of industries including agriculture. It is possible that in the not so distant future, RAS operations could become more automated than they are today. The main advantage of this would be to reduce labor costs. However, labor is not the main cost of RAS, which is instead

Chapter 8: Other Topics

fish feed. In contrast a lot of hydroponic operations are becoming more automated where labor is the biggest operational cost. Nonetheless, there is labor intensive fish handling tasks that may one day be best handled by a waterproof robot. For now though robotics does not play a large role in RAS.

Chapter 9: Summary

The goal of this book is to provide an outline of a comprehensive design process to build a robust, economical RAS. It also aims to point out important considerations and tradeoffs in making a variety of design decisions. There is no such thing as a perfect design and any existing design can be continuously improved and upgraded. Rarely are things done right the first time, and in fact a key step of the design process is to evaluate the final constructed system and look for future improvements.

Most of the design decisions boil down to a trade offs between capital expenditures (CAPEX), operational expenditures (OPEX), and water quality. A RAS can easily be designed to provide crystal clear water at high stocking densities for any production goal if enough money is available. One could simply invest in huge, state of the art water filtration systems that triple filters all of the water. However, the initial cost of the system would be astronomical as would be the day-to-day operating expenses. The sales price of the very happy and healthy fish that the system produced would not be nearly enough to cover expenses. Since one of the goals of a fish farm is to make money this type of system is not sustainable. Instead a RAS designer must be smart, clever, and knowledgeable so that they can construct a system for less money, that has lower operating expenses, while still maintaining adequate water quality for healthy fish.

Chapter 9: Summary

Future RAS technological development will find ways to lower initial capital costs while also lowering operational costs. Lower capital costs will eventually come from using pre-engineered and manufactured components, rather than entirely custom systems. Capital costs per kilogram production will also decrease as larger farms are constructed. Operational costs will decrease thanks to lower electricity costs from low-head designs and reduced labor costs from increased automation. Further cost savings will come as fish genetics improve, leading to lower FCRs, lower disease susceptibility, and higher growth rates.

The following appendices contain useful information that can be used during the planning stages of designing and budgeting a RAS. There are also links for additional RAS resources and a list of RAS design companies. Use the information in this book wisely and best of luck with all of your future RAS endeavors.

Appendix A. Aquaculture Design Process

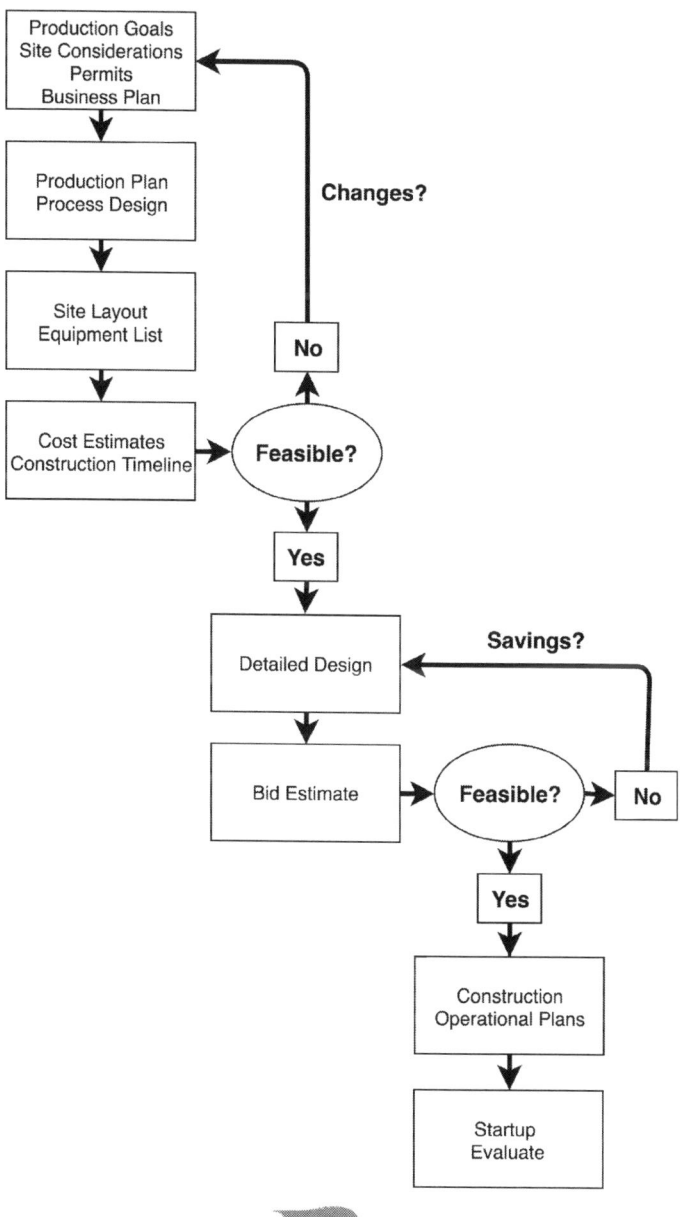

```
    ┌─────────────────────┐
    │  Production Goals   │
    │  Site Considerations│◄─────────┐
    │      Permits        │          │
    │   Business Plan     │          │
    └──────────┬──────────┘          │
               ▼                     │
    ┌─────────────────────┐          │
    │   Production Plan   │   Changes?
    │   Process Design    │          │
    └──────────┬──────────┘          │
               ▼                   ┌────┐
    ┌─────────────────────┐        │ No │
    │     Site Layout     │        └──▲─┘
    │   Equipment List    │           │
    └──────────┬──────────┘           │
               ▼                      │
    ┌─────────────────────┐      ╱─────────╲
    │   Cost Estimates    │─────▶│Feasible?│
    │Construction Timeline│      ╲────┬────╱
    └─────────────────────┘           │
                                      ▼
                                    ┌────┐
                                    │Yes │
                                    └──┬─┘
                                       ▼                  Savings?
                            ┌─────────────────┐      ┌───────────────┐
                            │  Detailed Design│◄─────┤               │
                            └────────┬────────┘      │               │
                                     ▼                                │
                            ┌─────────────────┐   ╱─────────╲      ┌────┐
                            │  Bid Estimate   │──▶│Feasible?│─────▶│ No │
                            └─────────────────┘   ╲────┬────╱      └────┘
                                                       ▼
                                                     ┌────┐
                                                     │Yes │
                                                     └──┬─┘
                                                        ▼
                                             ┌─────────────────┐
                                             │   Construction  │
                                             │ Operational Plans│
                                             └────────┬────────┘
                                                      ▼
                                             ┌─────────────────┐
                                             │     Startup     │
                                             │    Evaluate     │
                                             └─────────────────┘
```

Appendix B. Farm Outline

Fish Information
- Fish species
- Fish source
- Harvest size
- Initial size
- Growth rate
- Stocking density
- System temperature
- System salinity
- Disease susceptibility
- Feed load
- Feed size
- Feed type

Location Information
- Location
- Source water quality
- Source water flow rate
- Wastewater discharge
- Power supply & costs

System Basics
- System water volume
- Water flow rate
- Oxygen consumption
- Ammonia production
- Recirculation rate
- Water treatment methods
- Building size

Appendix B. Farm Outline

- Energy consumption

Planning
- Production goal
- Business plan
- Budget
- Equipment list
- Design drawings
- Engineering partner
- Construction partner
- Construction timeline
- Permits

Other Considerations
- Processing method
- Transportation
- Staffing
- Waste treatment system

Appendix C. List of Components for Costing Consideration

- Water treatment equipment
    - Tanks
    - Pumps
    - Solids filtration
    - Biofiltration
    - Carbon dioxide stripping
    - Oxygenation
    - Disinfection
- Controls (sensors, PLCs, controllers)
- Purge system
- Waste treatment and disposal
- Water supply treatment
- Ozone generators
- Oxygen generators
- Liquid oxygen back up
- Feeding system
- Fish handling system
- Fish processing equipment
- Back up generator
- Installation labor of water treatment equipment
- Electrical (supplies & labor)
- Plumbing (supplies & labor)
- Heating/cooling system (building & RAS)

Appendix C. List of Components for Costing Consideration

- General Concrete
- General Earthwork
- Equipment freight
- Design fee
- Project management fee
- Power service fee
- Land purchase
- Permits
- Building structures
- Farm start up costs (feed, fry, wages, etc.)

Appendix D. 100 Design and Operation Questions

General
1. Are the various treatment steps done in the correct order?
2. Are employees adequately trained to manage the system?
3. How much system water is wasted each day?
4. Was a new or existing building used?
5. Is there enough aisle and ceiling space to allow for material handling?
6. Is there safe access to all equipment?
7. Were the correct materials used for the amount of expected corrosion?
8. Are there operation instructions easily available?
9. What is the warranty for the system and equipment?
10. Do you have enough redundancy in each process step?

Tanks
11. What is the rotational velocity?
12. What is the flow out of the bottom drain in gallons per square foot?
13. What is the DO in various points in the tank?
14. Is there adequate head to gravity flow out of tanks?

Pipes
15. Is the water velocity too high or too low?
16. Is there room in the pipe to ensure adequate flow if fouling occurs?
17. Can the pipes be easily cleaned of fouling?
18. Are waste lines sloped to prevent clogging with solids?
19. Are there check valves to prevent flooding and siphoning?

Appendix D. 100 Design and Operation Questions

Pumps
20. Are the pumps corrosion resistant?
21. Are the pumps able to reach the desired flow?
22. Is there redundancy in critical flows?
23. Are pumps cycled/tested regularly?
24. How are pumps primed?
25. Is air entrained in the pump suction?
26. Do the pumps have to be greased regularly?
27. Can submersible pumps be lifted easily?
28. Are the pumps macerating any solids?
29. Are the motors TEFC and 1.15 duty rated?

Solids Filter
30. What is the drum micron size?
31. Is the drum filter large enough?
32. Is the drum filter located low enough in the water column?
33. Is there anything in the water that could clog the filter?
34. What is the pressure of the backwash pump?
35. Is the backwash water heated?
36. Is the backwash water system water or fresh water?
37. How often is the drum rotating and does it start and stop?
38. In case of clogging is there an area for overflow?
39. What is the water clarity and seki disc reading?
40. Can you remove a drum filter easily if needed?
41. If a drum filter goes down can the flow still be handled?

Fine Particles
42. Is ozone used and is the ORP measured?

Biofilter
43. Is there adequate surface area?
44. Is there solids accumulation in the biofilter?
45. Is an adequate flow rate continuously provided?

Appendix D. 100 Design and Operation Questions

46. What families of bacteria are present?
47. Is the bed properly fluidized?

Degassing
48. What kind aeration is used?
49. Is the building ventilated?
50. How is CO2 measured? Is it accurate?
51. Is CO2 below 10-12 mg/l?

Oxygenation
52. Is there adequate flow to the contactors?
53. What is the pressure of the contactor?
54. Does the oxygenated water mix evenly?
55. Is backpressure in the LHO slowing the flow rate?
56. What is the assumed oxygen consumption per kg of feed?

pH Management
57. What is added to replace alkalinity?
58. How much does the pH swing in a day?

Wastewater
59. How are nitrates removed?
60. How is phosphorous removed?
61. How are the solids dewatered?
62. Where does the wastewater go?

Source water
63. What is the flow rate?
64. Is there enough flow for future expansion plans?
65. Where does the water come from and what is the quality?
66. Is the flow and quality consistent?
67. Is the water hard?
68. Does the water have chloramine or chlorine in it?

Appendix D. 100 Design and Operation Questions

69. Does the water need to be treated?

Heating and Cooling
70. How is temperature maintained?
71. Are there copper heat exchangers involved in any process?
72. Is their independent heating and cooling for each system?

Disinfection
73. What is the UV dosage?
74. Are strict biosecurity protocols in place?
75. How much ozone is needed?
76. How is ozone mixed into the system?

Depuration/Purging
77. How long are fish purged?
78. How is Geosmin production being managed?

Water testing
79. How often is water tested? What are the parameters?

Fish Handling
80. How are fish crowded?
81. How are fish grated?
82. How are fish moved between tanks?
83. How often are fish grated/screened?

Fish Feeding
84. How are fines controlled?
85. How is feed and FCR tracked and managed?
86. Is the diet consistent?
87. Is the biomass carefully tracked?
88. Is photoperiod manipulation used at all?

Fish Supply
89. Is the fish source disease free?
90. How are fish acclimated to the system?

Controls
91. What is the level control?
92. Are the controls remotely monitored/accessed?
93. Are the probes regularly cleaned?
94. Are the probes reliable?

Backup systems
95. Is the generator exercised?
96. Is the emergency oxygen system present/adequate?
97. Is oxygen bled in the emergency oxygen system to prevent clogging?

Building
98. Are the building materials corrosion resistant for saltwater systems?
99. Is exposed concrete coated?

Economics
100. What are the price assumptions?

Appendix E. 1000 MT Farm by the Numbers

Building square footage	7,500 - 8,500 m²
Installed building cost	$5400/m²
Start up budget	$18-20 million ($20/Kg of production)
Water volume	7,100 - 10,000 m³
Water required	350-700 m³/day (5-10% replacement per day)
Fish feed per day	3.3 MT
Power required	475 Kw + heating/cooling
Operating budget	$5-6 million per year ($5-6/kg for salmon)

Appendix F. Water Treatment Equipment Choices

Equipment	Advantages	Disadvantages
Tanks		
Plastic	Inexpensive, no on site construction, cheap small tanks	Maximum size of 11 m^3, may require a stand
Fiberglass	Corrosion resistant, easily customized, cheap medium tanks	Minimum size of 11 m^3 gallons, onsite assembly, quality and lifespan are variable
Glass-coated steel	Easily shipped, UV resistant, recyclable, long lifespan, cheap medium-sized tanks	Minimum size of 11 m^3 gallons, onsite assembly, needs a concrete base
Concrete	Long life, cheap big tanks	Long construction time, variable costs, requires engineering
Solids Filtration		
Settling basin	Low-head, low-tech, no moving parts, no electricity	Requires a lot of space, size dictates efficiency
Radial flow separator	Low-head, low-tech, concentrates solids for removal, greater efficiency than settling basin	Size dictates efficiency, custom design is expensive, can only handle small to medium flow rates

Appendix F. Water Treatment Equipment Choices

Equipment	Advantages	Disadvantages
Parabolic filter	Low-head, low-tech, self-cleaning	Only good for small flow rates
Cartridge filter	Low-tech, inexpensive,	Only good for small flow rates, requires regular backwashing,
Bead filter	Removes small particles, doubles as a biofilter	High-head, requires regular backwashing, maximum flow rate of 10 m^3/h
Sand filter	Removes small particles, inexpensive	High-head, small to medium flow rates, can clog with high solids loading
Drum filter	Low-head, handles large flow rates, continuously backwashing, gold standard	Expensive, requires electricity, requires regular maintenance

Appendix F. Water Treatment Equipment Choices

Equipment	Advantages	Disadvantages
Biofilter		
Trickle filter	Inexpensive, dual use for CO_2 stripping,	Medium head, low specific surface area
Static bead	Simple, no moving pieces	Clogs easily, Requires backwashing, low specific surface area
Bead filter	Medium specific surface area, dual use for solids removal	High head, requires backwashing, max flow rate of 10m^3/h
Sand filter	High specific surface area, inexpensive media	Requires agitation with water, must be continuously operated
Microbead filter	High specific surface area, low head	Expensive custom design, requires agitation with water
Mixed bed bioreactor	High specific surface area, low head, reliable	Requires aeration, expensive media
Ion exchange	Instant start up, no bacteria needed	Media needs to be refreshed regularly
Carbon Dioxide Stripping		
Aeration tower	High gas to liquid ratio	High pumping head, need for custom towers
Surface aerators	No pumping head, medium gas to liquid ratio	Requires a basin, large spray radius
Subsurface aeration	No pumping head, small foot print	Low gas to liquid ratio, may require multiple passes, requires a basin

Appendix F. Water Treatment Equipment Choices

Equipment	Advantages	Disadvantages
Disinfection		
Ultraviolet	Reliable, easy to maintain, no residual is left in the water	Lightbulbs degrade over time, needs to be cleaned regularly
Ozone	Low-head, helps with fine particle filtration	Harmful at high concentrations in fish tanks, need large expensive generators, uses pure oxygen
Oxygenation		
Oxygen cone	High dissolution efficiency, small footprint, Off the shelf equipment, inexpensive	High head loss, only oxygenates a side stream, requires high pressure oxygen source
Low-head oxygenator	High dissolution efficiency, low-head	Low saturation concentration, custom design
U-tube	Medium dissolution efficiency, low-head	High construction costs, requires good soil, medium dissolution efficiency
Fine bubble diffusers	Inexpensive, no foot print, reliable	Low dissolution efficiency, requires high pressure oxygen

Appendix F. Water Treatment Equipment Choices

Equipment	Advantages	Disadvantages
Pumps		
Submerged axial flow	Large flow/low head, self-cooling, inexpensive	Hard to maintain, motor can flood, risk of electrocution, low-head only
Line-shaft	Large flow/low head, easily maintained	Expensive, low-head only
Submerged centrifugal	Medium flow/medium head, inexpensive, self-cooling	Inefficient at low heads, motor can flood, risk of electrocution
Centrifugal	Medium flow/medium head, inexpensive, easy to maintain	Inefficient at low heads, multiple pumps needed for large flows
Booster	Low flow/ high head, easy to maintain	Only efficient at high head, low flow
Fine Particle Filtration		
Foam fractionation	Safe and easy to operate, provides auxiliary CO_2 stripping	Not effective in freshwater, expensive equipment, requires extra pumping
Ozone	No extra pumping, helps with disinfection, works in freshwater	Needs to be used up, harmful at high concentrations, needs large expensive generators, uses pure oxygen

Appendix F. Water Treatment Equipment Choices

Equipment	Advantages	Disadvantages
Heating/Cooling		
Submerged electrical	Simple, no extra piping	Prone to overheating, high energy costs
Inline electrical	Gives precise temperature control	Water must be pumped through, high energy use
Natural gas boiler	Lower energy costs, low capital cost, can handle very high heating loads	Requires a heat exchanger, has its own separate piping network
Heat pump	High energy efficiency	High capital cost, large footprint
Air-cooled chiller	Energy efficient, no water source needed	Large footprint, must be outdoors
Water-cooled chiller	Small footprint, energy efficient	Requires a continuous water source
Alkalinity and pH Adjustment		
Baking Soda ($NaHCO_3$)	Cheap, non-caustic	Slow-acting
Calcium Carbonate ($CaCO_3$)	Cheap, non-caustic	Slow-acting, increases hardness
Calcium Hydroxide ($Ca(OH)_2$)	Fast acting	Caustic, increases hardness
Sodium Hydroxide ($NaOH$)	Fast acting	Caustic

Appendix G. Top Ten Most Common Mistakes

1. Bad water source

Mistake: Choosing a location with inadequate water flow, bad water quality, fluctuating water quality and flow, or expensive water costs.

Outcome: Additional water treatment equipment is needed or farm is incapable at operating at capacity.

Solution: Test water ahead of time, have multiple water sources, have more than enough water flow.

2. Poor equipment selection

Mistake: Using the wrong materials, inexpensive materials, or poorly manufactured equipment.

Outcome: Maintenance costs and equipment replacement costs become too high.

Solution: Purchase high quality, battle tested equipment upfront. Regularly schedule maintenance.

Appendix G. Top Ten Most Common Mistakes

3. Market swings

Mistake: Lack of anticipation for changes in final fish product wholesale price.

Outcome: Fish sales price is lower than needed to sustain farm operating expenses.

Solution: Properly study the market and current wholesale prices. Understand current supply of the product and how your farm will affect that supply. Assume a pessimistic number for the sales price.

4. Not enough operating cash

Mistake: Inadequately budget for operating costs while fish are growing and cash flow is zero.

Outcome: Inability to cover operating costs, or underinvesting in feed, labor, and equipment maintenance.

Solution: Have one to two years of operating cash on hand at farm start up.

5. Wrong fish species

Mistake: Select a fish species that has inadequate growth rate, no existing market, or poorly understood biological parameters.

Outcome: Low sales prices, high cost of production, or inadequate production.

Solution: Select a well-known species or perform pilot studies on a less known species.

Appendix G. Top Ten Most Common Mistakes

6. Adapting existing equipment or location

Mistake: Purchasing an old aquaculture operation or equipment from an existing operation and adapting the engineering and design to fit that location or equipment.

Outcome: High cost of equipment maintenance, and compromised system design that leads to high operating costs or inadequate filtration.

Solution: Invest in a new location and new equipment to start a farm.

7. Bad biosecurity

Mistake: Not taking proper biosecurity measures.

Outcome: Disease outbreak resulting in major fish losses, which leads to a stop in production and reduced revenue.

Solution: Invest in adequate disinfection equipment, purchase fish from a reputable vendor, and enforce proper biosecurity measures throughout the farm.

8. Poor management and staffing

Mistake: Hiring an inexperienced or unmotivated farm manager and staff.

Outcome: Staff does not take personal accountability for mistakes and/or lack of farm performance.

Appendix G. Top Ten Most Common Mistakes

Solution: Recruit and hire well qualified staff by offering competitive and wages and benefits, also tie farm performance to bonus compensation.

9. Inadequate or unproven system design

Mistake: Building the system with new, unproven filtration techniques and inflexibility to make needed changes.

Outcome: A system that underperforms and does not provide adequate water quality, reducing production and revenues.

Solution: Work with time-tested water filtration techniques and hire a trusted design engineer.

10. Regulations and permitting

Mistake: Not properly understanding local regulations or vetting possible bureaucratic hurdles.

Outcome: System construction is delayed significantly, system is shut down for violations, or compliance is too costly.

Solution: Operate in an area where other farms are present, and comply with all local regulations, plan for construction delays and work alongside the local permitting office to move things along.

Appendix H. Business Plan Outline

1. Executive Summary
    a. Production goals
    b. Species
    c. Market edge
    d. Industry outline
    e. Business objective

2. Company Summary
    a. Ownership group
    b. Engineering team
    c. Operating staff
    d. Retail partners

3. Products
    a. Live or processed
    b. Size of fillet
    c. Quality of fish

4. Market Analysis Summary
    a. Target market
    b. Competitors
    c. Long term price fluctuations
    d. Comparable fish species

Appendix H. Business Plan Outline

5. Strategy
   a. Offtake agreements
   b. Sales strategy- restaurants, grocery stores, direct sales
   c. Growth plans
   d. SWOT analysis

6. Financial plan
   a. Initial investment
   b. Profit and loss analysis
   c. ROI

Appendix I. Operations Checklist

Daily or weekly

- Visually examine the behavior of the fish
- Visually examine the water quality
- Check hydrodynamics in tanks
- Check distribution of feed from feeding machines
- Remove and record dead fish
- Wipe off membranes of oxygen probes
- Check water levels in sumps
- Check drum filter spray nozzles
- Record water chemistry tests
- Check oxygen cone or bead filter pressure
- Check pH and adjust
- Check UV light operation
- Check alarm system

Monthly

- Clean biofilters
- Check oxygen tank levels
- Calibrate pH-meter
- Calibrate feeders
- Calibrate Oxygen probes
- Run alarm tests
- Check emergency oxygen systems
- Check all pumps
- Check generators and test
- Grease bearings

Appendix I. Operations Checklist

- ☐ Rinse drum filter spray bar
- ☐ Check sumps for sludge

Bi-annually

- ☐ Clean UV and change lamps
- ☐ Change oil or airfilters in relevant equipment
- ☐ Clean biofilter
- ☐ Service oxygen probes
- ☐ Change drum filter spray bar nozzles
- ☐ Change filter plates in drum filter

Modified from: A Guide to Recirculation Aquaculture. 2015. Jacob Bregnabelle, FAO.

Appendix J. How to be a Great Farmer

Becoming a great farmer does not happen overnight, and there is a reason many families remain farmers for generations; it takes experience and learning from mistakes to become great. Humans seem naturally predisposed to farming, home gardeners and cattle ranchers alike know the feeling of satisfaction that comes with watching something grow thanks to your effort and investment. However, in order to make money and be considered a great farmer there are three areas that must be mastered: biology, technology, and business. Understanding each area is key to operating a successful farming venture.

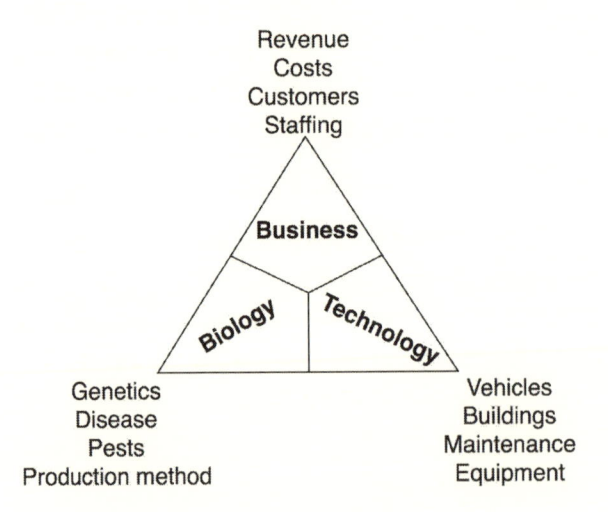

```
            Revenue
            Costs
            Customers
            Staffing
              /\
             /  \
            /Business\
           /----------\
          / Biology │ Technology \
         /_____|_____\

Genetics                    Vehicles
Disease                     Buildings
Pests                       Maintenance
Production method           Equipment
```

The first area, biology, is what most people think of when they think of farming. It involves understanding how to grow and manage the health of plants or animals. First the farmer must understand the genetics of their crop, whether that means purchasing hardy seeds or high quality calves. It means understanding how to prevent diseases and eliminate other pests. It also involves understanding the needs of the crop, how

Appendix J. How to be a Great Farmer

much water, what types of food, the right kind of soils, when to apply fertilizer, when it is ready to harvest etc. The great farmer controls every variable that he can to maximize yields.

The next area of expertise is operating and maintaining the technology that is required to operate a farm. That means purchasing and maintaining equipment such as vehicles, buildings, irrigation systems, and various other mechanical systems. A farmer needs to be able to fix things when they are broken and understand what tools are best for which jobs. These tools are what allow a farmer to maximize his productivity. Investing in technology and tools is key to competing in today's modern, globalized economy.

The final piece of being a great farmer is mastering the business side of the operation. This means tracking and managing costs, debts, assets, and revenue. It also means finding customers and hiring good people. Accounting, taxes, customer service, human resources, banking, and logistics are all part of the business side of farming. This piece is often overlooked, but if you can't run a business then you won't be able to run a farm.

Experience, knowledge, patience, flexibility, and attention to detail are all key traits that great farmers call on everyday to get the job done. There are no shortcuts or secrets, just hard work and persistence. Any farmer will tell you there are good years and bad years, things can be uncertain at times, but uncertainty can be managed.

Appendix K. Additional Resources and Reading

Books

Recirculating Aquaculture 3rd Edition. 2013. Michael Timmons and James Ebeling. Ithaca Publishing Company. http://a.co/brigykH

Aquaculture Engineering 2nd Edition. 2013. Odd-Ivar Lekang. John Wiley and Sons. http://a.co/iLCjqjN

Design and Operating Guide for Aquaculture Seawater Systems. 2002. J. Colt and J.E. Huguenin. Elsevier Science.

Handbook for Aquaculture Water Quality. 2014. Claude E. Boyd and Craig S. Tucker.

Aquacultural Engineering. 1978. Frederick Wheaton. John Wiley and Sons.

Water Quality: An Introduction. 2015. Claude Boyd. Springer

Web Resources

FAO: Guide to RAS www.fao.org/3/a-i4626e.pdf

Extension.org: Numerous PDF resources
www.articles.extension.org/pages/58711/recirculating-systems

World Aquaculture Society: Most active aquaculture society www.was.org

Freshwater Institute: Leading RAS research non-profit
www.conservationfund.org/our-work/freshwater-institute

Appendix K. Additional Resources and Reading

Aquaculture Journal: Leading scientific journal of aquaculture
www.journals.elsevier.com/aquaculture

Aquacultural Engineering Journal: Scientific journal with engineering emphasis www.journals.elsevier.com/aquacultural-engineering

Aquaculture Industry News:
www.undercurrentnews.com
www.intrafish.com
www.hatcheryinternational.com
www.aquaculturenorthamerica.com
www.aquaculturemag.com
www.thefishsite.com

Appendix L. List of System Designers and Equipment Providers

United States Companies
- Advanced Aquaculture Systems- FL www.advancedaquaculture.com
- Aquacare Environment- WA www.aquacare.com
- Aquaculture Systems Technology- LA/CO www.astfilters.com
- AquaDyne- TX www.aquadyne-filters.com
- Aqualogic- CA www.aqualogicinc.com
- Aquaneering- CA www.aquaneering.com
- Aquatic Equipment and Design- FL www.aquaticed.com
- Freshwater Institute- WV www.conservationfund.org/our-work/freshwater-institute
- HTH Aqua Group- GA www.hthaqua.com
- Integrated Aquaculture Systems- CA www.integrated-aqua.com
- Oceans Design- CA www.oceans-design.com
- Pentair Aquatic Ecosystems- FL www.pentairaes.com
- Pranger- IN www.prangerent.com
- WMT- LA www.w-m-t.com

International Companies
- Agrimarine- Canada www.agrimarine.com
- Akva Group- Norway www.akvagroup.com
- Aquabiotech Group- Malta www.aquabt.com

Appendix L. List of System Designers and Equipment Providers

- Aquacultur- Germany www.aquacultur.de
- AquaMoaf- Israel www.aquamoaf.com
- AquaOptima- Norway www.aquaoptima.com
- Aquatech Soultions- Denmark www.aquatec-solutions.com
- Argus- Canada www.arguscontrols.com
- Billund Aquaculture- Denmark www.billund-aqua.dk
- Canadian Aquaculture Systems- Canada www.canadianaquaculturesystems.com
- Fresh by Design- Australia www.freshbydesign.com.au
- HESY- Netherlands www.hesy.com
- Inter Aqua Advance- Denmark www.interaqua.dk
- JLH Consulting- Canada www.jlhconsulting.tv
- Llyn Aqua- UK www.llyn-aquaculture.co.uk
- MAT- Greece www.matlss.com
- Matsusaka- Japan www.matsusakaltd.co.jp
- Nofitech- Norway www.nofitech.no
- Kruger Kaldnes- Norway www.krugerkaldnes.no

Made in United States
Orlando, FL
13 July 2022